SPACE, TIME, AND MECHANICS

SYNTHESE LIBRARY

STUDIES IN EPISTEMOLOGY,
LOGIC, METHODOLOGY, AND PHILOSOPHY OF SCIENCE

Managing Editor:

JAAKKO HINTIKKA, *Florida State University, Tallahassee*

Editors:
DONALD DAVIDSON, *University of California*
GABRIËL NUCHELMANS, *University of Leyden*
WESLEY C. SALMON, *University of Pittsburgh*

VOLUME 163

SPACE, TIME, AND MECHANICS

Basic Structures of a Physical Theory

Edited by

D. MAYR

and

G. SÜSSMANN

Sektion Physik der Universität München

D. REIDEL PUBLISHING COMPANY

DORDRECHT : HOLLAND / BOSTON : U.S.A.

LONDON : ENGLAND

Library of Congress Cataloging in Publication Data
Main entry under title:

Space, time, and mechanics.

(Synthese library ; v. 163)
Papers from a symposium sponsored by the Deutsche Forschungs-
gemeinschaft.
Includes bibliographies and index.
Contents: Is the geometry of physical space a form of pure sensible
intuition? A technical reconstruction? Or a structure of reality? /
Günther Ludwig – Relations between the Galilei-invariant and the
Lorentz-invariant theories of collisions / Jürgen Ehlers – Geometrie und
Physik / C. F. von Weizsäcker – [etc.]
  1. Physics–Philosophy–Congresses. 2. Space and time–
Congresses. 3. Symmetry (Physics)–Congresses. I. Mayr, D.
(Dieter), 1944–   . II. Süssmann, Georg. III. Deutsche
Forschungsgemeinschaft.
QC5.56.S68   1983   530.1   82-21477
ISBN-13:978-94-009-7949-9   e-ISBN-13:978-94-009-7947-5
DOI: 10.1007/978-94-009-7947-5

Published by D. Reidel Publishing Company,
P.O. Box 17, 3300 AA Dordrecht, Holland.

Sold and distributed in the U.S.A. and Canada
by Kluwer Boston Inc.,
190 Old Derby Street, Hingham, MA 02043, U.S.A.

In all other countries, sold and distributed
by Kluwer Academic Publishers Group,
P.O. Box 322, 3300 AH Dordrecht, Holland.

D. Reidel Publishing Company is a member of the Kluwer Group

CONTENTS

PREFACE

In connection with the "Philosophy of Science" research
program conducted by the Deutsche Forschungsgemeinschaft
a colloquium was held in Munich from 18th to 20th May 1979.
This covered basic structures of physical theories, the
main emphasis being on the interrelation of space, time
and mechanics. The present volume contains contributions
and the results of the discussions. The papers are given
here in the same order of presentation as at the meeting.
     The development of these "basic structures of physical
theories" involved diverging trends arising from different
starting points in philosophy and physics. In order to obtain
a clear comparison between these schools of thought, it was
appropriate to concentrate discussion on geometry and
chronology as the common foundation of classical and quantum
mechanics. As a rather simple and well prepared field of
study, geochronometry seemed suited to analysing these
mutually exclusive positions.

D. Mayr and G. Süssmann (eds.), Space, Time, and Mechanics, vii.

ACKNOWLEDGEMENT

The editors gratefully appreciate the sponsorship of the
Deutsche Forschungsgemeinschaft and the cooperation of the
authors and publisher. It is also a pleasure to thank Frau
M.-L. Grohmann and Frau I. Thies for their organisational
and especially Frau B. Grund for typing and clerical work.

                    D. MAYR   G. SÜSSMANN

                    1982
                    University of Munich

INTRODUCTION

The distinct positions present at the symposium may be
roughly divided into three schools that differ in their
philosophical interpretation of physics and their metatheoretical conception of what a physical theory is or
should be: (1) The Constructivism of the Erlanger Protophysics (P. Lorenzen, P. Janich), (2) the Structuralism
within the Analytical Philosophy of Science (P. Suppes,
J.D. Sneed, W. Stegmüller), and (3) the Marburg Physics
Theory (G. Ludwig).
   (1) As a protagonist of constructivism, the "second
Erlanger program" tries to give a normative foundation of
the sciences, including physics, which originates from
every-day experience (Lorenzen and Schwemmer 1975, Janich
1980). Its aim is to adopt an immediate, theory-independent
approach which leads from the prescientific experience of
civilized mankind with handicrafts, to a normal language
and conceptual system covering basic notions of science.
The norms necessary to characterize the basic concepts are
required to initiate prescriptions and manual procedures
which deliver artifactual realisations of these very norms,
which all include standard etalons and gauges. The intention
is to establish constructively all the fundamental concepts
of lenght, duration, inertial mass and electric charge.
The protophysical concepts of action, experience and norm
serve as indisputably given basic terms which are assumed
to have a theory-independent, prescientific adequacy. This
prephysical program aims at a non-circular and purposive
foundation of physics in the framework of <u>Protophysics</u>
that - surprisingly enough - turned out to be antirelativistic.
Consequently, protophysical statements are supposed to be
valid beyond all scientific experience. In its own view,
Protophysics professes to be a certain kind of formal
pretheory for every physical theory, in which basic concepts
may be operationally defined, and manufacturing norms should
uniquely determine the construction of measuring instruments.
   Recently , however, more substantial aspects have arisen

ix

D. Mayr and G. Süssmann (eds.), Space, Time, and Mechanics, ix–xv.
Copyright © 1983 by D. Reidel Publishing Company.

in the protophysical analysis of real physical theories.
Within the very basic concept of a process, Protophysics
now seems to distinguish between <u>physical courses</u> and
purpose-directed <u>human operations.</u> Physical courses have
features that are not influenced by human actions, but are
uniquely determined by natural regularities. There are thus
situations in which the handicrafts performance of a protophysical norm depends not only on the individual ability of
the craftsman but also on the compatibility with some real
structure of the environment. This plausible statement is
based on the well known physical fact, that it is not
usually true that "matter can be compelled to adapt itself
to ideal norms" (Lorenzen und Schwemmer p. 236). Consequently,
the possibility or validity of norms is restricted by physical
experience. At the same time a remarkable mitigation of the
antirelativistic position was proposed by its co-originator,
P. Janich - at least in the case of special relativity -
which now seems compatible with a normative foundation.

(2) The second school, the structuralism of the
analytical philosophy of science, may be called an informal
conception relative to the formal approach of the so-called
Statement-View (Stegmüller 1979, §1). Carnap and the other
protagonists of this older philosophy tried to get a rational
reconstruction of scientific theories by means of a formal
language and an axiomatic set theory. This formalistic
approach, however, is burdened with extensive complications
which already arise within the formal reconstruction of
mathematics. As we have known since Frege, Peano, Russel
and Whitehead, and Hilbert and Skolem, even small proofs
in the theory of sets are so profuse in formal language,
that the foundations of more complex branches are hardly
workable in this style. But if the formalized text of the
syntax is enlarged by the introduction of new notions and
additional rules, the resulting language is more manageable.
This is an essential feature of, for example, Bourbaki's
presentation, in which formalized language has been condensed
to a more ordinary level, the usual language of all mathematical
texts in practice (Bourbaki 1968). In this manner Bourbaki
constructed a rather rigorous fondation for most of
contemporary mathematics. This could serve as an example and
as a tool for the philosophy of physics.
        In the fifties, P. Suppes started with usual or ordinary
language of mathematics to reconstruct the mathematical part
of a physical theory (Suppes 1957). Without particular

elaboration of the formalized language and set theory
(cf. Scheibe 1981), he introduced so-called set theoretical
predicates, which in any example are defined by a vocabulary
of condensed mathematical notions, yielding his informal-
syntactical approach. With these predicates, the mathematical
part of a physical theory is represented by a class of
structures which are finite tuples of sets, relations and
functions. Later, Suppes' approach was enlarged by J.D. Sneed
with so-called informal-semantical methods (Sneed 1971). In
this more "model theoretic" view Sneed separates two levels,
corresponding to a distinction between empirical and theoretical domains (called 'non-theoretical' and 'theoretical'
in relation to the given theory). These domains are represented by classes of so-called partial possible models
(M_{pp}) and possible models (M_p), which again are characterized
by set theoretical predicates. In the empirical domain, the
members of M_{pp} contain only structures (sets, relations)
defined and physically clarified in predecessor theories.
The essential feature of the empirical part is a certain
subclass of M_{pp} - the intended applications - whose important
members may correspond to paradigmatic examples and crucial
experiments of the physical theory. In the domain of the
possible models new concepts are introduced by "theoretical"
relations. Roughly speaking, they describe the new physical
aspects of the theory relative to its pretheories, i.e.
its genuinely novel features. Those members of M_p which
obey the physical laws are called models of the theory.
The two domains are connected by a projection from M_p to
M_{pp} which simply omits the theoretical relations. Finally,
certain theories combine to yield a theory tree connected
by the intertheoretical order relation of model theoretic
inclusion which represents a kind of superstructure on the
class of all theories (net-structure). This informal
structuralistic view "allows a constructive criticism and
a partial vindication of the philosophies of Kuhn and
Lakatos" (Stegmüller 1979, p. 14).

   (3) In contrast to the philosophical attempts mentioned,
G. Ludwig's Physics Theory has been developed from physical
problems - a metatheoretical by-product, as it were, derived
from the axiomatization of quantum theory (Ludwig 1970). As
is well known, the treatment of foundational questions in
physics often produces rough drafts of metatheories. These
physical answers to the philosophical question "What is
physics?" have a long tradition including such famous
exponents as Aristotle, Galileo, Newton, Leibniz, Mach, Duhem,

Poincaré, Einstein and Heisenberg.

An essential part of Physics Theory's mathematical dressing was fashioned by Nicolas Bourbaki, a pseudonym for a group of young French mathematicians who started their famous "Eléments de Mathématique" in the winter term of 1934/35. In the more than 20 volumes which have appeared to date, Bourbaki consistently employed the so-called axiomatic method and created something like the Euclid of the 20th century. This foundation of mathematics sums up the pioneering efforts of many celebrated mathematicians. For example, the works of Riemann, Dedekind, Cantor, Poincaré, Hilbert, Fréchet, Brouwer, Hausdorff and others were necessary to obtain that concept of a topological space which we can find in the lucidly and elegantly codified form of Bourbaki. The idea of the axiomatic foundation and presentation is simple. A mathematical object is not conceived by an explicit construction and not described by an ad hoc procedure peculiar to its specific nature; instead, it is presented as the combined result of a number of more general features or structures, each of which may also be found in other objects. The concrete concept is produced as a synthesis of more abstract notions. For example, the real line is defined as the commutative number field which has a continuous ordering. Explicit constructions are only needed for proofs of existence, being the self-consistency of the implicit definition. Some properties of the real number are thus already treated in the more general domains of algebraic (especially group theoretical) structures, others within the context of topological structures, or among order theoretical structures. The very combination of these parental qualities yields, to be sure, some novel features peculiar to the concept of a real number. But the axiomatic architecture leads to a much deeper understanding of the relations between the various mathematical disciplines, and they are presented in a unified form. According to Bourbaki, a mathematical theory is a species of structures: a relation, typified on the echelon above some principal and auxiliary sets, together with transportable axioms (Bourbaki 1968, IV, § 1.4). On this basis, special insights may easily be transported to other mathematical branches. Mathematics has been given a quite remarkable boost by the systematic use of the axiomatic method as practised on a grand scale by Bourbaki.

Furthermore, what is even more important in our context, Bourbaki helped to clarify what contemporary mathematics,

with its remarkable standard of rigor, really is. The
philosophy of physics is well advised to accept this formal
paradigm. We need not, for example, discard the law of the
excluded middle, or the axiom of choice, as most construct-
ivists would have us believe. A mathematical "axiom" or
postulate is nothing but part of the nominal definition of
an abstract structure which does not need any concrete
interpretation. Mathematics is independent of physics and
is prior to all laws. The notion of geometry should thus
be split into mathematical geometry as a descendant of
topology, and physical geometry together with chronometry
as the zeroth chapter of mechanics.

In this respect, Ludwig's Physics Theory may be regarded
as a program for physics, somewhat analogous to that of
Bourbaki's for mathematics. Clearly, unlike in theoretical
mathematics, physical concepts are submitted to empirical
interpretation, and they need additional characterization
describing the correspondence to real objects and experimental
results.

The first step of Ludwig therefore takes the following
direction. Each well formulated physical theory PT consists
of three parts: a mathematical theory MT, a domain of
reality W, and a correspondence (——) between MT and W; in
short, we have the identification PT = MT(——)W. Here, MT
is a theory in the sense of Bourbaki, a structure which is
richer than the theory of sets (an appropriate species of
structures). Yet the empirical correspondence is not a
mapping; it is more like a many-many relation because, in
general, certain elements of MT are related to various
physical states. This blurred kind of assignment reflects
the typically physical situation where the mathematical
results of PT do not tally exactly with the corresponding
experimental results.

The second step allows for the fact that some physical
concepts cannot be defined independently of PT since the
physical content of PT is not completely exhausted by the
components MT, (——) and W. For example, the force fields
of electrodynamics, the dissipation notions of thermodynamics,
and the state concepts of quantum theory are theoretical
relations which can only be given physical interpretations
if the whole physical theory is used. This foundational
problem of physical concepts, the so-called problem of
theoretical terms, is answered in Ludwig's view by restriction of the domain of reality W to a subdomain. This subset
G of the given facts has to be selected in such a way that

the physical concepts of MT(——)G can be defined independently
of PT, and that the theoretical content of PT, corresponding
to the domain of reality, may already be built up with
MT(——)G.

We are not going to present a detailed representation of
Ludwig's Physics Theory (cf. Ludwig 1978 or the short account
of the L-program in Hartkämper and Schmidt 1981). Let us just
mention two aspects which may indicate the physical adequacy
of Physics Theory.

Imprecisions are a fundamental fact of life in physics:
Results of measurements hold only within a certain range of
accuracy, which precludes the empirical correspondence from
becoming an exact mapping. Moreover, a mathematical theory
trying to picture reality is not a precise representation of
the world, and we cannot single out one approximating scheme.
In Ludwig's approach the various concepts of imprecision are
described by one mathematical tool, the uniform structure
imposed on the base sets of MT. By canonical extensions of
the uniform structure to power and product sets, the defined
relations of PT are endowed with uniformities which may
characterize the properties of the imprecisions. In addition,
this imprecision structure plays a fundamental role in the
comparison of theories, particularly in the case of
approximative reduction and embedding, so typical of all
theoretical physics.

In the second aspect we return to the problem of the
foundation of physical concepts. Is it possible, by the
axiomatic method of structure species, to ensure a physical
interpretation of theoretical relations? Physics Theory offers
a simple criterion for solving this problem. Let us assume
that in MT the (principally) undefined base sets and structures
are interpreted by known physical concepts; for example,
they may coincide with the interpreted sets and structures
of a pretheory. Such a form of MT is called the axiomatic
base of PT. (It is certainly not a trivial task to find,
for an arbitrary PT, a physically equivalent form which
satisfies the conditions of an axiomatic base.) Now, a new
structure deduced in (the axiomatic base representation of)
MT will get a physical interpretation. It is uniquely
determined by the interpreted sets and relations which
are used for the deduction. If there are certain interpreted
relations of MT and corresponding measurements which can be
combined to yield an indirect measurement of the deduced
structure, it is reasonable to say that the structure
represents a set of real physical facts. This may justify

calling it an established <u>physical structure</u> or a new physical concept (cf. Ludwig 1978, § 10.5).

REFERENCES

Bourbaki, N.: 1968, 'Theory of Sets', Hermann, Paris.

Hartkämper, A. and H.-J. Schmidt: 1981, 'Structure and Approximation in Physical Theories', Plenum Press, New York.

Janich, P.: 1980, 'Die Protophysik der Zeit', Suhrkamp Verlag, Frankfurt am Main.

Lorenzen, P. und O. Schwemmer: ²1975, 'Konstruktive Logik, Ethik und Wissenschaftstheorie', BI Mannheim.

Ludwig, G.: 1970, 'Deutung des Begriffs "physikalische Theorie" und axiomatische Grundlegung der Hilbertraum-struktur der Quantenmechanik durch Hauptsätze des Messen', Lecture Notes in Physics, 4, Springer (in addition: 'Foundation of Quantum Mechanics', 2volumes, Translation in Englisch in preparation, to appear by Springer).

Ludwig, G.: 1978, 'Die Grundstrukturen einer physikalischen Theorie', Springer.

Scheibe, E.: 1981, 'A Comparison of Two Recent Views on Theories' in Hartkämper and Schmidt 1981.

Sneed, J.D.: 1971, 'The Logical Structure of Mathematical Physics', Reidel, Dordrecht.

Stegmüller, W.: 1979, 'The Structuralistic View of Theories', Springer.

Suppes, P.: 1957, 'Introduction to Logic', D.v. Nostrand, Princeton.

Günther Ludwig

IS THE GEOMETRY OF PHYSICAL SPACE A FORM OF PURE SENSIBLE INTUITION? A TECHNICAL CONSTRUCTION? OR A STRUCTURE OF REALITY?

In this paper we shall not be able to present a complete or definitive answer to the above question. We only attempt to examine this question without prejudice and we shall not make the claim that only one of the above three possibilities is correct. In fact, we find that each of these viewpoints has a certain justification. We shall attempt to clarify this problem by examining the relationships between these three viewpoints.

We wish to confine the problem to the case in which Euclidean geometry can be used to describe the geometry of the real space – that is – we shall not consider problems concerning "cosmology" or "black holes".

§ 1   Three extreme viewpoints

In my opinion each of following three extreme viewpoints I to III is incorrect.

I) Euclidean geometry is nothing other than a form of pure sensible intuition which is a "necessary basic requirement" for all physics. We leave open the question how human beings may have gained this form of pure sensible intuition, whether this form was gained during evolution by natural selection or was gained in the first years of our life, or is a form impressed in our mind. As a "necessary basic requirement" we mean that Euclidean geometry will be a necessary structure for the formulation of experience.

The following two arguments can be raised in opposition to this opinion:

1) In the history of physics we have found that it is possible to use non-Euclidean geometries in physics. On the other hand it is not possible to see how we are able by pure sensible intuition to state whether a physical realization of a plane – obtained for example by means of a "grinding procedure of three plates" – corresponds to an intuitive plane. In other words: How are we be able to determine whether a geo-

1

D. Mayr and G. Süssmann (eds.), Space, Time, and Mechanics, 1–20.
Copyright © 1983 *by D. Reidel Publishing Company.*

metric figure obtained by technical procedures corresponds to
our intuitive notions.

2) The discussion of problems in the foundations of mathematics has weakened our faith in intuition. We no longer believe that it is possible to verify the axioms of Euclidean
geometry by intuition - that is - we are not convinced by intuition that Euclidean geometry is internally consistent. For
example we are not convinced that there can be only one line
passing through a given point which is parallel to a given
line. By intuition we cannot exclude the possibility that
there exists a finite range of angles (see fig. 1) for straight

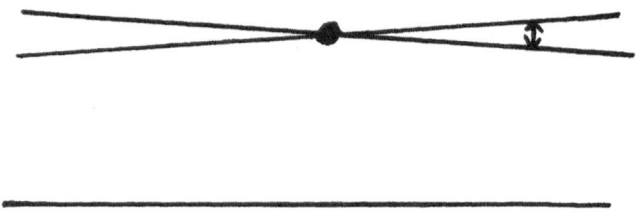

Fig. 1

lines passing through a given point which does not intersect
a given straight line. This angle may depend upon the distance between the given line and the given point.
        If we examine our understanding of geometrical figures it
is not difficult to see that we have developed a certain intuition which is not sufficiently precise that one may deduce precise mathematical relationships. For example we have
no intuition for the behaviour of straight lines, planes,
etc. at infinity. Our intuition corresponds to the case in
which only a bounded portion of geometry can be seen, and
the remaining portion cannot be observed.

II. Another extreme opinion is that Euclidean geometry is a
technical constructed structure which is used to obtain
a reference system for the description of physical phenomena
and that such constructions are arbitrary as - for instance -
the well known Euclidean maps of our earth. Such maps are
very useful. Nevertheless the Euclidean geometry of the map

is not identical with the geometry of the surface of the earth. In physics we are not interested in arbitrary maps of the space but in the actual ("physical") geometry of the space regardless of how the actual geometry is determined - from experience or by other methods.

III. The third extreme opinion is as follows: The theorems of Euclidean geometry may be found in nature, the theorems can be "read" from nature despite the fact that we know that the mathematical form of the theorems includes idealizations about nature. In this sense one believes Euclidean geometry to be a structure which is present in nature.

It is evidently no simple matter to read the Euclidean geometry from nature. Realizations of geometrical figures such as straight lines or planes cannot be found in nature, they have to be "made". But how do we make such geometrical figures?

If we reflect on these three extreme viewpoints it seems that it is possible to raise objections to each of these viewpoints by means of the others. Therefore we have to conclude that each of these viewpoints is right in some respect and we are wrong if we claim that one of these is the complete solution of the space problem.

§ 2   Methods and interests

We may obtain a better insight into the problem if we admit that various persons have various interests concerning the space problem and that these various interests will determine the methods used to describe and solve the problem, or more precisely, to solve the parts of the problem of interest.

For example the method depends upon whether one is asking what has to be anticipated before every experience with objects existing side by side; or whether one is asking, what are the necessary postulates on manipulation of solid bodies for the purpose of measuring distances between spots (finite physical realizations of points) in the real space; or whether one is asking, what are the "real structures" of space which make it "possible" to construct measuring apparatuses in the usual way.

Since our primary interest is in this last question, we shall not consider questions concerning the historical development of physics. We only seek the formulation of a physical theory by which it is possible to answer the question: what aspects of geometry describe a real structure?

To find a better methodological procedure to answer such
questions as "what is real" and "what is possible" we have
tried to develop a special formulation of a physical theory in
(G. Ludwig (1978)) and have improved this formulation in
(G. Ludwig (1981)). I have called this formulation an axiomatic basis.

§ 3 An axiomatic basis for a theory of space

To obtain an axiomatic basis for the theory of the physical
space that is - to elaborate the real structure of space - we
have to avoid the use of such measured quantities as the distances between several spots in the space. We will also not use
technical instructions to produce special geometrical figures
as planes or straight lines. We wish to restrict the basic
domain of the theory (basic domain = Grundbereich, defined in
(G. Ludwig (1978)), to qualitative relations between solid
bodies which we can recognize without any theory. Such qualitative relations are essential in order to put together a
puzzle. Such relations are that the various pieces can be
transported and fitted together. We will not go in the details
of the description of such relations. We wish only to mention,
that "protophysics" also uses such relations.
        H.J. Schmidt has demonstrated (H.J. Schmidt (1979)), that
it is possible to develop an axiomatic basis of a space theory using only such qualitative relations as the fundamental
relations. In this theory Euclidean geometry is obtained as
a deduced structure. We then "say", that we have detected Euclidean geometry as the real structure of the physical space,
at least locally - that is - in the area in which we can work
with solid bodies in the desired manner. But what is the
meaning of such a statement? We have the impression that this
statement is often misunderstood.
        First we do not claim that it is possible to recognize
immediately such a structure in nature - that is to "read"
(approximately) in nature theorems of Euclidean geometry as
described above in III. "Immediately" means that we use no
theory. On the contrary we claim that, given a mathematical
structure in the theory, this structure reflects something
(at least approximately) of reality. In this context it is
essential that we look upon various but equivalent mathematical structures as the "same"structure which differs only in
the form of description. In this sense "Euclidean geometry"
need not be described by the form of Euclid's axioms. The
form of various equivalent structures has nothing to do with

reality.

Another misunderstanding states that geometrical figures which can be defined in the mathematical theory (e.g. planes or straight lines) exist in nature without human actions. To claim that mathematical objects in a mathematical theory (as a part of a physical theory) are also real objects is a deep misunderstanding of the nature of a physical theory. Let $\mathcal{MT}$ be the mathematical theory as a part of a physical theory. Then $\mathcal{MT}$ by itself does not say anything concerning the question: What are the real facts? On the contrary the real facts have to be added to $\mathcal{MT}$ in the language of $\mathcal{MT}$. In (G. Ludwig (1978)) I have designated this theory $\mathcal{MT}$ with the added facts as $\mathcal{MTA}$. The statement that the mathematical structure established in $\mathcal{MT}$ is a picture of a real structure is an abbreviation for the following situation:

1) If we add facts (including those which are produced by human actions) to $\mathcal{MT}$ and obtain the theories $\mathcal{MTA}$ in this way then these theories $\mathcal{MTA}$ prove to be self-consistant. We then say that the facts are consistent with the theory.

2) If the physical theory is closed in a sense I have described in (G. Ludwig (1978)) § 10.3 then from $\mathcal{MTA}$ it is possible to conclude (not only from $\mathcal{MT}$!) new <u>real</u> facts and - what is essential - new <u>possible</u> facts. To emphasize that these concepts of "real" and "possible" have at first a special "physical" interpretation I have called these concepts in (G. Ludwig (1978)) "physically real" and "physically possible". It is permissible to go beyond this "physical" interpretation and believe that these physically real facts are also ontologically real.

There are two different contexts in which we may have physically possible facts. These can also be mixed. In the first context are those facts which we have "at our disposal" - those which may be produced if we wish. In the other context are those facts which must be described in terms of probabilities (see (G. Ludwig (1978)) § 11).

Concerning our theory of space the preceeding discussion has the following significance:

1.) The experiences with solid bodies are consistent with the theory of space if we make "undisturbed" experiments. We will later discuss what we mean here by "undisturbed".

2.) We can make solid bodies of the various geometrical forms -

that is - of such forms which are "physically possible". Such
"formed" solid bodies can be for example - a cupboard manufactured of wood or a statue. As a real structure of space
Euclidean geometry means: We should be able to at least determine in advance what form of solid bodies we are able to
produce and transport - for example - to shift a drawer in a
cupboard.

The investigation of "physical possibilities" in the
context of an axiomatic basis of the structure of space also
includes the question of the possibility to measure distances.
The distance between points is a structure which can be deduced in the mathematical theory we described in (H.J.Schmidt
(1979). A measurement of such a deduced structure is what we
have called a indirect measurement in (G.Ludwig (1978)) § 10.3.
The distance is not introduced in the framework of the axiomatic basis by an operative definition. This structure is obtained by mathematical deductions. This yields several "physical possibilities" to measure indirectly this distance.
One of these possibilities is the measurement by means of
chains consisting of solid bodies as proven by H.J. Schmidt
(1979).

This procedure in the framework of the axiomatic basis
does not prohibit us (in another methodological construction
of a theory of space) to "define" distance in an operational
way - for example by means of chains as described in (G.Lud-
wig (1974-1979)II. Such an operative method has the disadvantage, that it is very difficult to maintain our awareness of
the presuppositions concerning the manipulation of solid bodies
which are necessary to obtain an exact definition.

§ 4   Relations between the axiomatic basis and protophysics

The last considerations give us the possibility to discuss
the relation of the axiomatic basis of a theory of space to
the normative construction procedure proposed by protophysics
(see for instance (P. Janich(1976,1980)).

The following remarks cannot give a general and complete
elaboration of a comparison between the protophysical construction of the geometry (P.Janich (1976,1980)) and the
construction of an axiomatic basis of geometry (H.J.Schmidt
(1979)). The following remarks can only be an invitation to
make such a comparison between the two procedures and to make
this comparison precisely using mathematical structures. This
seems to be possible since there is also a precise mathematical description of the protophysical construction of geometry,

Since the two procedures result in the Euclidean geometry as mathematical picture there can be no differences in
the mathematical picture. As we have already mentioned above
and as described in more details in (G.Ludwig (1978)) two
equivalent mathematical theories (also if developed in different forms) constitute the same theory if there are no principle differences in the interpretation of the mathematical
picture. Therefore possible differences of the two theories
can be found only in the interpretation. Are there such
differences on principle in the interpretation? If yes, what
are the differences?

If one looks superficially on the two theories the difference seems to be the following. The axiomatic basis describes
geometry as a "real structure" of the world. In contrast to
this opinion protophysics establishes geometry "a priori" by
"norms" as a condition to start physics at all. If I understand not only the procedure of the axiomatic basis but also
a little bit of protophysics the situation seems to me to be
more complicated.

Since I understand the procedure of the axiomatic basis
much better than the procedure of protophysics I will try to
discuss the protophysical foundation of geometry relative to
the axiomatic basis and not vice versa.

Protophysics emphasizes the fact that the norms can not
be disproven by experience because the norms are prescriptions for making and for the aims of making. Norms can only
be disproved by logical conclusions in a similar fashion as
in mathematics. Nevertheless protophysics lays stress on the
fact that there has to be a rich source of prescientific experience - experience obtained from handicraft activities.
Without this source of experience it would be impossible to
formulate norms as prescriptions for manipulation of solid
bodies or as aims of such manipulations. If, for instance,
in the protophysics of solid bodies, we speak of shifting
such bodies and of contact between two such bodies it is
presupposed that this makes sense and that we know what we
are speaking about.

The basic relations of the theory $\mathcal{PJ}_1$ (the so-called pre-
theory to $\mathcal{PJ}_2$ in (H.J.Schmidt (1979)) by which the interpretation of the total axiomatic basis of the space theory
is established) are interpreted by concepts already known
from handicraft activities. Such a concept is the transport
of solid bodies (a concept on which the concept of the transport of space regions is founded in (H.J. Schmidt (1979)).
The transport of solid bodies is certainly one of the most

essential basic experiences and one of the fundamental processes of handicraft activities. In addition other concepts
for example - two solid bodies are in contact or that one
body fits in a holding device - are very similar in both procedures, that of the axiomatic basis and that of protophysics.

Therefore it is my opinion that there in principle are
no differences between protophysics and the axiomatic basis
if we examine only the first basic relations. The only differences in the basic relations is that the method of axiomatic basis takes the meaning of these basic relations for
granted and that protophysics has investigated and analysed
the structure of this region of experiences obtained from
handicraft activities. This investigation is certainly a
valuable contribution of protophysics. The method of the
development of a physical theory in terms of an axiomatic
basis has taken this "pre"-scientific region for granted
because this method is only concerned with the structure of
physical theories.

The conformity between protophysics and axiomatic basis
in the pre-scientific region does not imply that the domain
of solid bodies for which the relations and axioms of (H.J.
Schmidt(1979)) apply is the same as the region of those
bodies to which the actions normed by protophysics apply.

The requirements for a physical theory described in
(G.Ludwig (1978)) do not include anything concerning the
methodology to be used to find the axioms. The requirements
only state how we are to test the theory using experiments
belonging to the basic domain ("Grundbereich" (G. Ludwig
(1978)). The axioms can not be read in nature, on the contrary they sometimes determine (together with other characteristics) the basic domain. In this sense some of the
axioms in an axiomatic basis may be "selection rules" for
the "intended" experiments.

Another valuable contribution of protophysics is that it
lays particular stress on the fact that there are many relations and facts which are not read in nature but are produced on the basis of aims which we wish to reach. May it
perhaps be possible that protophysics has gone too far in
this direction?

In an axiomatic basis of the space theory it is also
possible that some axioms may be "selection rules" for the
"interesting" facts. The axiomatic basis gives no other prescriptions how we may achieve the desired "selection". Protophysics describes these "selections" in more detail by
rules for manipulation and desired objectives - that is by

norms. In the case of the space theory the axioms of $\mathcal{P}\mathcal{J}_1$ in (H.J. Schmidt (1979)) are more of the character of selection rules for the "undisturbed" experiments with solid bodies. The axioms of $\mathcal{P}\mathcal{J}_1$ in (H.J. Schmidt (1979)) are introduced to obtain the concepts of "space region" and of "transports" of space regions by an equivalence relation which permits us to disregard the special properties of the composition of the solid bodies. These axioms ones formulated say in the sense of an axiomatic basis that it is "physically possible" to manipulate solid bodies in such a way, that no contradiction to the axioms arises.

The axiomatic basis does not describe the way to reach such an "undisturbed" manipulation. The notion, that these axioms are "laws of nature" is only an abbrevation for the statement that this so-called undisturbed manipulation is "physically possible".

If we look at protophysics from this view the two methods seem to be compatible relative to the situation described above. And in addition; the two methods have some similarity in this "selection of the undisturbed handling" with solid bodies if we do not at first look on those norms by which special geometric figures may be produced. The undisturbed manipulation in the sense of the axiomatic basis is nothing other than the postulate, that "shapes" ("Gestalten" in the sense of protophysics) are reproducible and that this reproducibility is uniquely determined if we try to manipulate the solid bodies in conformity with the basic axioms and if we try to obtain results which do not depend on the composition of the bodies.

Therefore I dare to state that protophysics and the axiomatic basis are related to the "same" region of "undisturbed"manipulation with solid bodies. I know that there can be objections made by protophysics. We shall discuss these objections later.

If this compatibility and similarity of the two methods could be based more precisely, where then are the differences between protophysics and the axiomatic bases of space theory? These differences arise in two ways and are due to the concepts of physical real and physically possible (as introduced in (G.Ludwig (1978)).

To discuss the differences we assume that protophysics and the theory $\mathcal{P}\mathcal{J}_1$ of (H.J. Schmidt (1979) are compatible as just described – that is – with respect to what $\mathcal{P}\mathcal{J}_1$ calls solid bodies and transports of these bodies and what protophysics calls solid bodies and to be in contact and to shift

one body against another and so on but <u>without</u> the norms for
the constructions of special geometrical figures.

A consequence of the axiomatic basis of (H.J. Schmidt
(1979)) is that the "undisturbed" experiments with solid bodies are <u>uniquely</u> defined by the axioms which permit us to
introduce the space regions as equivalence classes. The methods described in (G.Ludwig (1978)) say that in this case
the space regions are "physically real" - that is - if one
has an "undisturbed" solid body this body determines uniquely
a space region, which is independent of the special solid
body used to determine experimentally the space region.

Naturally a theory cannot logically determine what experiments look like. It would be logically conceivable that
there is more than only one possibility to define space regions (independent of special properties of the composition
of the solid bodies) without violating the axioms of $\mathcal{P}\widetilde{\mathcal{T}}_1$
- that is - that there are more than one possibility to de-
fine what are "undisturbed" experiments with solid bodies.
If there would be more than one such possibility then the
theory $\mathcal{P}\widetilde{\mathcal{T}}_1$ of (H.J. Schmidt (1979)) would not be defeated
by experiments. But there would be many different theories
with the same mathematical picture having different basic
domain which exclude one another. The different basic domains
would determine the various different possibilities to introduce space regions. In such a case a new theory has to be
found, which comprises all these possibilities. The new theory would then (in the sense of (G.Ludwig (1978))tell us,
that space regions are not "physically real" but "physically
possible" and the new theory would specify these various
physical possibilities.

If there is only one possibility of definitions of undisturbed experiments with solid bodies by the axioms of
$\mathcal{P}\widetilde{\mathcal{T}}_1$ and the postulate of independence of the composition
then there is the conjecture that already <u>some</u> of the norms
of protophysics (without the norms for special geometrical
figures!) determine the "undisturbed" (in the sense of protophysics) manipulation with solid bodies. We do not wish
to decide this question at the present time. It could be
that protophysics actually needs also some of the norms for
special geometrical figures to uniquely determine the "undisturbed" manipulation with solid bodies. We want to discuss the theory of (H.J. Schmidt (1979)) in relation to some
other norms of the protophysics.

If $\mathcal{P}\widetilde{\mathcal{T}}_1$ and protophysics are concerned with the same
region of manipulation of solid bodies then it is possible

to translate the norms of protophysics in the mathematical language of $\mathcal{PT}_1$ and thus of $\mathcal{PT}_2$. In this manner it is possible to formulate without difficulaty the final result of a grinding procedure of three plates as a relation in $\mathcal{PT}_2$. We can ask whether this relation is in contradiction to the theory $\mathcal{PT}_2$. It is not difficult to see that this relation can not be in contradiction to the theory $\mathcal{PT}_2$ since $\mathcal{PT}_2$ describes an Euclidean geometry. It may be similar with the other norms of protophysics. Nevertheless there is again a fundamental difference between $\mathcal{PT}_2$ and protophysics if protophysics insist on the statement that also the construction of geometrical figures has to be normed.

In the axiomatic basis the additional axioms of $\mathcal{PT}_2$ are regarded as "laws of nature" – that is – as structure laws governing the space regions and their transports which have already been uniquely introduced. On the contrary the protophysic looks on the geometrical figures as added by norms – that is – these figures are not given by experiences but have to be constructed as prescribed by the norms.

Of course the differences between these two opinions can not be seen in the experiments and by handling with solid bodies as long as the theory $\mathcal{PT}_2$ of (H.J. Schmidt (1979)) describes an Euclidean geometry since in this case $\mathcal{PT}_2$ states that geometrical figures postulated by the norms of protophysics are "physically possible" – that is – the norms can be "realized". The difference of the two opinions can be seen if there exist difficulties to "realize" the norms. Here the word "realize" is used in the sense of (G. Ludwig (1978))- that is – in the sense of $\mathcal{PT}_2$ and not in the sense of protophysics. It is easy to encounter misunderstandings if we do not distinguish the various meanings of "realize" in (G. Ludwig (1978)) and in the context of protophysics.

In the context of protophysics the word "realize" is used for the statement of a handicraft procedure what must give the desired result if it is possible to apply the procedure in a concrete situation. For instance the grinding procedure of three plates is such a "realization" for a plane surface if it is possible to apply this procedure – that is – to grind, to come at an end of grinding, and so on. In the context of (G.Ludwig (1978)) and therefore of $\mathcal{PT}_2$ the word "realize" is used if one has produced concrete facts which were "physically possible". For instance, plane surfaces of three plates fitting one another are "physically possible" in $\mathcal{PT}_2$. Three concrete plates together with their postulated properties are "physically real" facts in $\mathcal{PT}_2$ and

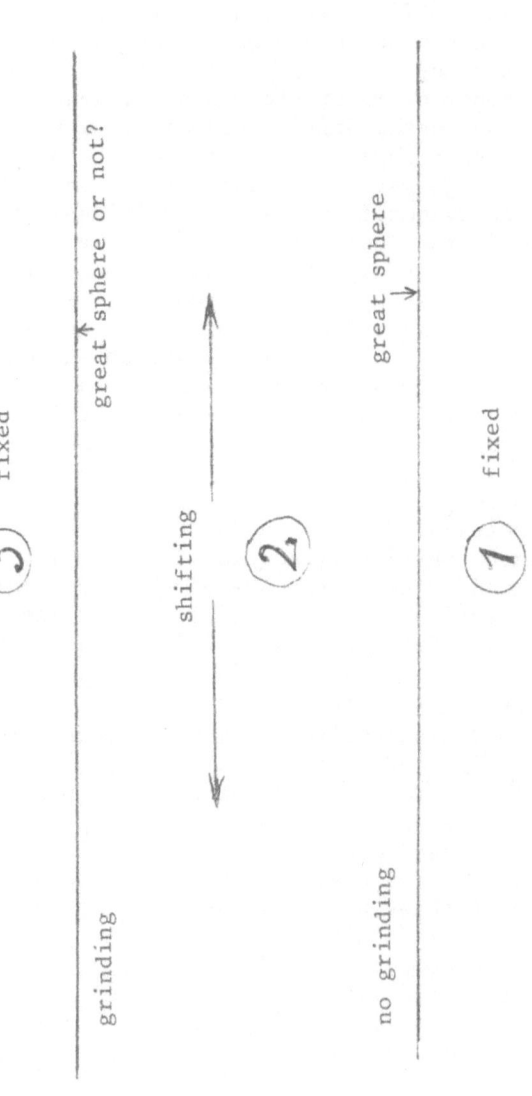

Fig. 2

inthis sense a "realization" of a (physically) possibility. $\mathcal{PT_2}$ can not give any procedure how to realize a (physically) possibility!

The first axioms of (H.J. Schmidt (1979)) which permit us to introduce the concepts of space regions and their transports do not determine the geometry as an Euclidean one – that is – one can change other axioms of $\mathcal{PT_2}$ without contradiction in the mathematical theory, for instance one can introduce such axioms that the geometry is that of a uniformly curved space (see (H.J. Schmidt (1979)). Because of the mobility of the solid bodies postulated by axioms the grinding procedure of three plates is also possible in these uniformly curved spaces and will give "great spheres" (Großkugeln). We will therefore look at another experiment which in the sense of (H.J. Schmidt (1979)) turns out differently in the case of an Euclidean or a uniformly curved space. We take three solid bodies shortly named 1,2,3. 1 and 2 may be two bodies of a grinding procedure of three plates – that is – the contact surface of the two bodies 1 and 2 (see fig. 2) is a piece of a plane in the case of Euclidean geometry or a piece of a "great sphere" (Großkugel) in the case of a uniformly curved space. Now the bodies 1 and 3 may be fixed and 2 may be shifted against 1 and 3. Between 1 and 2 may be a film of oil stopping any new grinding between 1 and 2. Between 2 and 3 a grinding will take place until a new contact surface is generated in such a manner that the body 2 can be shifted against 3 along this surface (see fig. 2).

If the space is Euclidean then the surface between 2 and 3 will be a piece of a plane – that is – will satisfy the norms of a grinding procedure of three plates. If the space is uniformly curved then the surface between 2 and 3 will not be a "great sphere" – that is – will not satisfy the norms of a grinding procedure of three plates. And the discrepancy of the surface between 2 and 3 and a "great sphere" will be so mucht the more as the body 2 is more and more "thick".

If the experiment will turn out as described in the second case then in the sense of (G. Ludwig (1978)) we have stated that the geometry of the real space is not Euclidean. In the sense of protophysics this experiment should be called "disturbed" and rightly so since "disturbance" is defined in protophysics relative to the norms.

But what is to be done in the sense of protophysics if the experiment turns out as described? One has to try to remove the "disturbances". However if the opinion of the method

of the axiomatic basis is right that the basic axioms uniquely define the space regions and their transports and in
this sense the undisturbed experiments then there would be
no possibility to remove the "disturbances" postulated by
protophysics. Every attempts to remove these disturbances
would necessary results in violating other norms - that is -
in introducing other "disturbances".

We see that again the essential question arises whether
the basic axioms of (H.J. Schmidt (1979)) uniquely determine
space regions and their transports. How shall we find the
answer to such a question? Mathematics and the theory in itself can not give the answer as already described above.
Only experiences can give the answer. One has to look for
other possibilities of experimentation to satisfy the basic
axioms of (H.J. Schmidt (1979)) - that is - one has to look
for more then one technical construction of a geometry in
space. To look for more than one geometry is nothing else
than to look for disturbances in the sense of protophysic.
The norms of the protophysics are then in this sense a prescription how to select the "right" geometry satisfying "all"
the norms.

Why are we convinced that there is only one possibility
to introduce space regions and their transports? Just since
until now we could not find any other possibility to require
the desired equivalence relation be independent of the properties of the composition of the solid bodies.

And what would we have to do if perhaps we would later
find other possibilities to introduce space regions and
transports? This would not be a "revolution" in physics.
Nothing would collapse. Only a new knowledge of real structures in the world would arise and a new more comprehensive
theory would have to be found. For instance for such a form
there would be more then one geometry of the space. (For the
concept of a more comprehensive theory see (G. Ludwig (1978)).

Provided that there is only one possibility to introduce space regions and their transports we guess that protophysics uses more norms than necessary - that is - that fewer
norms would suffice to determine the "undisturbed" handling
with solid bodies.

There is another point of differences between protophysics and axiomatic basis. This difference concerns the physical significance of the basic axioms of $\mathcal{PT}_1$ in (H.J.
Schmidt(1979)), of the axioms which permits us to define the
essential equivalence relations and to introduce by means of
these relations the concepts of space region and their trans-

ports. These axioms are called laws of nature. But why? It
is not to deny that at first the handicraft manipulations of
solid bodies does not by itself comply with the basic axioms
of $\mathcal{P}\mathcal{J}_1$ in (H.J. Schmidt (1979)).For instance, heating of one
of the bodies would lead to other results and would lead to
contradictions to the axioms. If such heating takes place
then we had left the basic domain in the sense of the axiomatic basis or we had a "disturbance" in the sense of proto-
physic.The axioms of the axiomatic basis state that it is "physically possible" to manipulate solid bodies without contra-
dicting the axioms. The axioms do not say how to make such
a manipulation. They say only that it is possible to eliminate all "disturbances".

Protophysics would say that we would have to abandon the
norms if we are unsuccessful for all our efforts. The physicists are very interested in formulating that which is successful. This formulation is found by the axiomatic basis. We will
now give examples which will show that the question whether
we are successful or not is not trivial.

As first example we shall try to construct a geometry in
a carrousel (roundabout).According to the basic axioms of
$\mathcal{P}\mathcal{J}_1$ or the corresponding norms we have be successful in the
neighbourhood of the axis of the rotation. But as we go far
away from the axis we obtain insurmountable difficulties.

If we try to manufacture measuring apparatuses in the
carrousel according to the norms far away from the axis we
are unsuccessful and if we try to take such measuring apparatuses which are manufactured in the laboratory according
to the norms, in the carrousel far away from the axis then
the apparatuses are destroyed. In the sense of the protophysics the construction of a geometry is only successful in
the neighbourhood of the axis. In the sense of the axiomatic
basis a space geometry exists approximately only in the neighbourhood of the axix and does not exist (is "physically impossible") far away from the axis.

Fortunately there are many possible ways to construct
geometries as reference systems according to either the
norms of protophysics or according to the axioms of $\mathcal{P}\mathcal{J}_1$.
I have called such systems "nearly-inertial" systems (see
(G. Ludwig (1974-1979) II). If one raises the standards of
precision to satisfy the norms of protophysics or resp. to
satisfy the relations in $\mathcal{P}\mathcal{J}_1$ then the reference systems
thus obtained are more restricted to those systems which one
has been previously called inertial systems.

For the manufacturing of measuring devices according to

the norms of protophysics it is necessary to be successful
only locally in finite regions. For the application of the
theory $\mathcal{P}\mathcal{T}_\gamma$ it suffices that the theory is a good description of the experiments only locally and in finite regions. As
well the "ideations" ("Ideationen" see (P. Janich (1976,
1980)) in the protophysics as the "idealizations" in (H.J.
Schmidt(1979)) ("Idealisierung" see (G. Ludwig (1978)) need
not to be of any experimental significance. We need only
"small" measuring devices produced in our laboratories or
factories to start the development of physics.

      It would be imaginable without any influence on our
"small" measuring devices that "big" solid bodies can not
be transported at all. What shall mean this statement? An
example: In a factory may be manufactured a very big body
(a bridge or something similar). This body may be composed
of several "small" parts. To transport the big body to the
desired position it will be decomposed in his parts. The
parts may be transported to the desired position and we try
to compose the parts at this new position. Will we succeed?
We have the experience to succeed. But why do we succeed?
It would be imaginable that we would not succeed if the bodies are sufficiently large.

      In such situations protophysics would require us to look
for causes for the "disturbances" defined by the deviation
from the norms. Then protophysics would postulate the eli-
mination of these causes in order to obtain the undisturbed
behaviour. The method of the axiomatic basis would identify
the "real geometry" of space as the "cause" for the following reason. The undisturbed transport of "small" solid bodies determines small space regions and their transports.
The perhaps empirical fact that "big" regions can not be
transported is a real structure of the space since it can
not be avoided without violating the axioms for the definition and transports of "small" regions. Since in practice
very large space regions (if we do not take space regions
of cosmic dimensions) may be transported the axiomatic basis
interprets this fact as a "real structure" of the space.

      Our reflections demonstrate that protophysics would
not lead to constructions of Euclidean maps (see point II
at the beginning) of the space if the space would be non
Euclidean in the sense of (G. Ludwig (1978)). In such a
situation protophysics would make demands to the engineer
which he cannot accomplish in principle, which he can not
"realize" (in the sense of (G.Ludwig (1978)). Protophysics
would perhaps not say that the demands of the norms can not

be accomplished in principle but only that one has not succeeded <u>until now</u> to work according the norms. But in any case these far reaching demands for "big" bodies need not to be accomplished if we want to start and to develop physics. It suffices to work with "small" bodies.

There are objections to the concepts of "unrealizable", resp. "physically impossible" or "to be physically excluded" (see the extensive discussion of these concepts in (G. Ludwig (1978)). If we would exclude such concepts then we would have to give up big parts of physics even such parts which are very essential for technology - for example - thermodynamics where the "impossibility" of a perpetuo mobile of the second kind plays a central role as a law of nature.

Here we wish to make a remark concerning some objections to the statement that geometry describes a real structure of the space, objections which are of totally different kind than those of protophysics. To discuss these objections we assume that it was stated by experiments, that "big" solid bodies can not be transported. The objection to the interpretation that the real space has a non-Euclidean structure is the following. It is possible to use an Euclidean geometry (for instance a special map in the sense of point TT at the beginning) <u>and</u> to introduce a field of force of such a kind, that "<u>small</u>" bodies are deformed (relative to the Euclidean geometry) independent of their composition but that "big" bodies are deformed in a way which dependends on their composition. Such an "explanation" of the deformations relative to the Euclidean geometry by forces is a total misunderstanding of the aim of an axiomatic basis as founded in (G. Ludwig (1978)). The species of structure of a non-Euclidean geometry used in an axiomatic basis, where "small" space regions can be transported without deformations, can be <u>represented</u> in another mathematical theory, for instance, in an Euclidean geometry where the Euclidean geometry is a special map of the non-Euclidean space. The Euclidean structure of the map has no physical real significance as can be seen by the methods of (G. Ludwig (1978)§ 10) and follows already from the fact that there are many different Euclidean maps with corresponding different fields of force. Euclidean geometry plus corresponding field of forces is only another form of the description of the <u>same</u> structure namely the real structure of a non-Euclidean geometry (see in general to such questions the developments in (G. Ludwig (1978) § 7).

Let us summarize the result of our discussion of the

relations between protophysics and the method of an axiomatic
basis. The protophysical construction of special geometry
frames seems to be compatible with the axiomatic basis in
(H.J. Schmidt (1979)).

The axiomatic basis tries to elaborate the relation between theory and reality, for instance to see in what sense
the distance is a description of real structure. Therefore
the axiomatic basis neglects to see for experimental methods
with which one may "realize" (in the sense of (G.Ludwig (1978))
what the axiomatic basis tells us to be "physically possible".
The axiomatic basis does not deliberately provide special
methods how to measure indirectly what is physically real, for
instance how to measure the "physical real" distances. The
experimental physicists are invited to invent such methods.
The axiomatic basis does also not give any special instructions or prescriptions for the manipulating with solid bodies to avoid disturbances - that is - to get the uniquely
defined structure of space regions and their transports.

Protophysics emphasizes the analysis of handicraft pre-
experiences and of all such aims for actions which have to
be pre-supposed if one wants to start physics. In this sense
the protophysic can be a valuable supplement to the axiomatic
basis.

Protophysics introduces geometrical concepts by concrete
instructions and prescriptions and gets closer to the historical development of physics than the method of axiomatic basis.

It is my opinion that one problem in the procedure of
Protophysics has not yet been sufficient clarified. Perhaps
the method of the axiomatic basis could be helpfull to clear
this problem. It seems to me that protophysics uses "too many"
norms. It could be that fewer norms are sufficient to establish the physical geometry.

The problem whether one has introduced "too many" norms
has two aspects.

At the one hand it is possible that one has introduced
norms which already follows "logically" from other ones. This
problem can be solved mathematically.

At the other hand - and this is the more important aspect - it can be that some of the norms (not necessary all)
uniquely determine how we may manipulate solid bodies without disturbance. One has in reality no possibility to fulfill the next part of the norms if these norms are not accomplished automatically "without disturbance" - that is - by
fulfilling the first part of the norms. This second problem
is much more difficult than the first one since it can not

be solved mathematically.

This second problem characterizes exactly - that is my opinion - this point where at the moment are the main diffe- rences (not only concerning the space problem) between pro- tophysics and the methods of (G. Ludwig (1978)).

§ 5 Axiomatic basis, protophysic and sensible intuition.

Let us proceed now to ask another question concerning the first viewpoint. What about the relationship between pure sensible intuition and the axiomatic basis of a physical theo- ry of space? Is the notion of pure sensible intuition per- haps unnecessary for physics or even useless or does it only have historical relevance? For the axiomatic basis it was assumed that we are capable of manipulating solid bodies in such a way that we can solve the problem of a puzzle. This manipulation is only possible if we have a capability to see the shapes of the bodies what we are told by the so-called "gestalt psychology".

In order to state the axioms in an axiomatic basis it is necessary to formulate our intuition for all prescientific experience with solid bodies. It seems that the imposition of such a requirement in the development of an axiomatic basis is far too complicated to be able to be solved by means of a pure sensible intuition description of Euclidean geometry. Here we note that, in the beginning of this paper we have already found that such a formulation is highly questionable. Therefore the following interesting question remains: What are the presuppositions which make it possible for children to play with bricks without knowing any "geometry"

The following question arises for protophysics: How are the norms invented? Without intuition it seems that it would be impossible for a craftsman to invent plans for his work. It remains questionable whether an intuitive insight into Euclidean geometry is necessary for such plans.

One structure of Euclidean geometry is neither a pure sensible intuition nor the result of a normative manipula- tion nor a real structure: it is the structure of the in- finity of the Euclidean space.

The infinity of the Euclidean space can not be imagined intuitively as we have already mentioned at the beginning.

The infinity can not be constructed since it is impossib- le to make infinite manipulations.

The infinity in the mathematical theory of the axio- matic basis cannot be an insight in a real structure since

it is possible to test the mathematical picture only by finite bodies and finitely many bodies.

But what is the origin of that structure of infinity in
the Euclidean geometry?

This infinity is in the case of an axiomatic basis only
a representative of our ignorance to what distances the Euclidean geometry is a good picture of reality. Our ignorance
is compensated in the mathematical picture by an idealized
extrapolation, the extrapolation of an infinite space. Our
knowledge can change during the development of physics. Such
changes lead to new more comprehensive theories. These more
comprehensive theories allow to describe the older theories
as approximations (see (G. Ludwig (1978) § 8) . The methods
described in (G. Ludwig (1978)) do not prescribe how to
introduce idealizations to replace ignorance. All possibilities of idealizations are allowed if these idealizations
produce a handy mathematical theory.

In protophysics is more claimed for its "ideations"
(see (P.Janich (1976,1980)). The result is the same as that
of (H.J. Schmidt (1979)).

A physicist will regard these "idealizations" or "ideations" as provisional - that is - as fictious parts which
have no real and no practical significance for physics.

I would like to thank Prof. Carl A. Hein, Mathematics
Department of Stonehill College for his valuable assistance
in preparing the english language version of this paper.

REFERENCES

Inhetveen, R.: 1981, ' Norm und Form. Eine konstruktive
    Begründung der Euklidischen Geometrie'(Habilitations-
    schrift Erlangen).
Janich, P.: 1980, 'Zur Protophysik des Raumes; in : Böhme,G.
    (editor) Protophysik, Frankfurt.
Janich, P.: 1980, ' Zur Protophysik der Zeit', Suhrkamp
    Verlag.
Ludwig, G.: 1978, 'Die Grundstrukturen einer physikalischen
    Theorie', Springer Verlag.
Ludwig, G.: 1974-1979, 'Einführung in die Grundlagen der
    theoretischen Physik', Vieweg-Verlag.
Ludwig, G.: 1981, 'Axiomatische Basis einer physikalischen
    Theorie und theoretische Begriffe', Zeitschrift für all-
    gemeine Wissenschaftstheorie XII/1, 55-74.
Schmidt,H.J.: 1979, 'Axiomatic Characterization of Physical
    Geometry',Lecture Notes in Physics 111, Springer-Verlag.

Jürgen Ehlers

RELATIONS BETWEEN THE GALILEI-INVARIANT AND THE LORENTZINVARIANT THEORIES OF COLLISIONS

1. INTRODUCTION

The purpose of this paper is to describe, by means of a simple example, some relations between two physical theories of which the older one is considered as being contained, as an approximation, in the later one. We consider the so-called non-relativistic theory of collisions between particles, due to Wallis, Wren and Huyghens, which is based on the Galileo-Newtonian kinematics, and its successor, the special-relativistic collision theory due to Lewis and Tolman, based on the kinematics of Lorentz, Poincare, Einstein and Minkowski.

In section 2 a general kinematical framework is laid down which contains an unspecified, non-negative parameter of dimension (speed)$^{-2}$. The values $\varepsilon=0$ and $\varepsilon>0$ correspond, respectively, to Galilean and Einsteinian kinematics. In the latter case $\varepsilon^{-1/2}=c$ is the fundamental speed the coincidence of which with the speed of light in vacuo is of no relevance here. Then, in section 3, a general collision theory is constructed. It is based on the general kinematics and on an energy conservation law which is assumed to have one and the same form in all inertial frames of reference. The mere existence of one such conservation law is shown to determine the functional dependence of the energy on the speed and to imply a law of conservation of linear momentum. Besides linear momentum the concepts of proper and inertial mass, and of internal and kinetic energy are defined without specification of the value of the parameter ε. Thus the common conceptual and formal structure of the collision theories based on Galilean and Einsteinian kinematics, respectively, is made explicit before those propositions are considered by which the cases $\varepsilon=0$ and $\varepsilon>0$ differ.

The most important qualitative difference is that in the Galilean case inertial and proper mass coincide and this speed-independent mass is additively conserved, whereas in the Einsteinian case the product "inertial mass $\times c^2$" is an energy function (and thus conserved), while in general

21

D. Mayr and G. Süssmann (eds.), Space, Time, and Mechanics, 21–37.
Copyright © 1983 by D. Reidel Publishing Company.

proper mass is not. In the Einsteinian case the particular
energy function given by the famous equation $E = m_{inertial} \cdot c^2$
is the only one which is the time-component of a (conserved)
world-vector valued function, the energy momentum. (In this
theory it is, therefore, convenient to reserve the term
"energy" for the values of this preferred energy function,
to use "mass" instead of the cumbersome "proper mass", and
to drop the term "inertial mass" altogether.) In the Newtonian case mass and momentum form a world vector, and mass,
momentum and energy form a "5-vector", i.e. they transform
according to an indecomposable five-dimensional representation of the Galilean group.

In section 4 it is stated in which sense the Galilean
theory is contained, as an approximation, in the Einsteinian
one, and some remarks concerning this approximation relation
form the end of this paper.

The several statements just referred to, except the one
concerning the mass-momentum-energy five-vector, are well
known, and also the essence of the logical arrangement presented in this paper has already been communicated some time
ago (references are given below), but it appears to me that
this presentation has not been used as a standard example to
illustrate the relation between two successive theories. This
example is intended to show in particular that not only the
quantitative laws, but also the concepts and interpretation
rules of the two theories can be related to each other by
the device of reconstructing them as special cases of a single
parameter-dependent theory. This method of exhibiting one
theory as a "limit" of another one can be applied also to
the more complicated case of Newton's and Einstein's theories
of gravity (see Friedrichs 1927, Dautcourt 1964, Künzle 1976,
Ehlers 1981) and perhaps to the still more complicated relation between classical and quantum mechanics. (See, e.g.,
Kundt 1966, Hepp 1974, Scheibe 1981 and, in particular,
Ashtekar 1980.)

The example considered below also shows how strongly the
propositions which follow from a conservation law and relativity principle, depend on the assumed spacetime structure or
symmetry group, so that experimental tests of these propositions may be considered as indirect tests of that spacetime
structure.

In this paper I do not take into account "massless" particles such as photons. Their inclusion would require a
slightly more general concept of state than the one employed
here, and a corresponding extension of the collision laws

$(C_1)-(C_4)$ in section 3.

2. GENERAL KINEMATICS

Let the geometry of three-dimensional Euclidean space be
accepted as the physical theory of measurements of length
and determination of positions with respect to suitable bodies
of reference such as the Earth, as well as one-dimensional
affine space as the theory of time and duration. Also, let
there be defined a relation of simultaneity relative to any
body B of reference, for example by means of slow transport
of clocks or some equivalent procedure (see, e.g., Rindler
1980, sections 2.17 and 2.6, footnote 8, p. 31). Then any
event can be labeled, with respect to each frame of reference, by four coordinates (t,x^ν) $(\nu=1,2,3)$ or by a position
vector $\vec{x}$ and a time t, as usual.

We assume in addition the existence of inertial frames
of reference S, S' etc. with respect to which free particles
move uniformly on straight lines. (The problem of defining
free particles is not considered here. (See, however, the
excellent analysis by R.A. Coleman and H. Korte (1981) which
applies to any theory based on a differential spacetime manifold.) For the purpose of collision theory it suffices that
the particles have well-defined velocity vectors $\vec{v}_1,\vec{v}_2,\ldots$
immediately before and after the collisions. In fact, the
spacetime structure is used only locally, in the vicinity of
each collision event; only for simplicity it is formulated
globally.) Moreover, we assume that if two free particles
have equal velocity vectors in one inertial reference frame
S, they have equal velocity vectors in any other such frame.
Finally we suppose that the same units of length and duration
are employed in all inertial frames, that the ratio between
the period of a clock moving uniformly relative to S, measured
by means of clocks at rest in S, to the proper period of
that clock depends only on its speed relative to S, and that
a similar statement holds for measuring rods.

Clearly, all these assumptions are valid both in Galilean and Einsteinian kinematics. They are meaningful if only a
compact part of R^4 is considered as physically accessible.
Combining, e.g., well-known arguments of Einstein (1905),
Weyl (1923) and Rindler (1979) one deduces that the coordi-
nate-transformations $(t,x^\nu) \to (t',x^{\nu'})$ relating inertial coordinates form a subgroup G_ϵ of the affine group of R^4. ϵ is
the non-negative, universal constant of kinematics already

mentioned. G_0 is the inhomogeneous Galilean group; G_ε with $\varepsilon>0$ is the Poincarégroup. Many methods can be used to measure the value of ε and thus to determine the correct kinematics within this general framework. Measurement of the time-dilation for a single clock-speed, the propagation of light, several electromagnetic phenomena or the laws of collision (see below) may be employed, as discussed by Rindler (1979), e.g. The coincidence of the results of such measurements of ε and the compatibility and simplicity of mechanical, optical, ... theories based on Einstein's kinematics - compared, e.g. with earlier ether - theories - establish that kinematics as an improvement of the earlier one.

The structure of space time can be described by means of the concepts of linear algebra and geometry as follows (compare Weyl 1923, Künzle 1972, Ehlers 1981):

The set of events is represented by the four-dimensional, real, affine (point-) space A^4. Let V^4 denote the corresponding vector space of translations or 4-vectors; it can be identified with the tangent spaces of A^4 considered as a manifold. A^4 is equipped with a time-metric and a space-metric. The former is defined by a 2-covariant, symmetric tensor g of signature $(+1,-sgn\varepsilon, -sgn\varepsilon, -sgn\varepsilon)$, the latter by a 2-contravariant, symmetric tensor h of signature $(-sgn\varepsilon, +1, +1, +1)$, where $sgn0=0$ and $sgn\varepsilon=1$ if $\varepsilon>0$. h and g are related by

$$h \cdot g = - \varepsilon 1 \tag{1}$$

or in index notation

$$h^{ab}g_{bc} = -\varepsilon \, \delta^a_c . \tag{1'}$$

g defines an inner product on V^4 whereas h defines such a product on the dual space $(V^4)^*$, the space of 1-forms. A spacetime vector X is called timelike if $g(X,X) > 0$; $(g(X,X))^{1/2}$ is its duration. Y is said to be spacelike if it is contained in the range of h, i.e. if there is a covector η such that $Y=h \cdot \eta$, and $h(\eta,\eta) \geq 0$; $|Y| = (h(\eta,\eta))^{1/2}$ is its length. In a system of inertia g and h are represented by the matrices

$$g = \begin{pmatrix} 1 & 0 \\ \hline 0 & -\varepsilon 1_3 \end{pmatrix}, \quad h = \begin{pmatrix} -\varepsilon & 0 \\ \hline 0 & 1_3 \end{pmatrix} . \tag{2}$$

If $\varepsilon > 0$, g is Lorentzian and $-h = \varepsilon g^{-1}$. If $\varepsilon = 0$, h is a (degenerate) Galilean metric and

$$g = dt \otimes dt \qquad (3)$$

where t is Newton's absolute time. In this case dt spans the kernel of h.

In any case the transformations of the group G_ε can be reinterpreted ("actively") as maps of A^4 leaving g and h invariant.

If ε tends to zero the light cone $\{Z \mid g(Z,Z) = 0\}$ degenerates into a hyperplane. The identity-component G_ε^o of the homogeneous subgroup of G_ε is simple if $\varepsilon > 0$; G_0^o is not even semi-simple. This is one mathematical reason why Einstein's kinematics is "simpler" than Galileo's (see also Künzle 1972).

The set V^3 of future-directed ($X^t > 0$) timelike unit vectors represents, according to the law of inertia, the translation-states of free particles. It has the structure of a three-dimensional Riemannian space of constant curvature $-\text{sgn}\varepsilon$. Its homogeneity reflects the fact that there is no preferred state of motion.

For later applications a kinematical lemma will now be stated. For this purpose two continuous, real functions (Strictly speaking, the values of T are squares of speeds, not real numbers) of two variables ε, v are defined in the domain given by the inequalities $0 \le \varepsilon$, $0 \le v$, $0 < \sqrt{\varepsilon} v < 1$ by

$$\gamma_\varepsilon(v) = \begin{cases} 1 & \text{if } \varepsilon = 0, \\ (1 - \varepsilon v^2)^{-1/2} & \text{if } \varepsilon > 0 \end{cases} \qquad (4)$$

and

$$T_\varepsilon(v) = \begin{cases} \tfrac{1}{2} v^2 & \text{if } \varepsilon = 0, \\ \varepsilon^{-1}(\gamma_\varepsilon(v) - 1) & \text{if } \varepsilon > 0. \end{cases} \qquad (5)$$

Straightforward calculation shows the validity of the following (See Fig. 1 below.)

lemma (Penrose and Rindler 1965; Ehlers, Penrose and Rindler 1965). If four free particles have the velocities $\vec{v}_1, \vec{v}_2 = -\vec{v}_1$, $\vec{v}_3, \vec{v}_4 = -\vec{v}_3$ with $v_1 = v_3$ and $\vec{v}_1$ orthogonal to $\vec{v}_3$ relative to an inertial frame S, and if the inertial frame S' moves with respect to S with any velocity parallel to $\vec{v}_1$, then the speeds of these particles relative to S' satisfy the equation

$$T_\varepsilon(v_3') = T_\varepsilon(v_4') = \frac{1}{2}(T_\varepsilon(v_1') + T_\varepsilon(v_2')). \tag{6}$$

3.   COLLISION THEORY

Consider particles moving uniformly on straight lines except during short intervals of time when some particles approach each other closely and change their velocities, and possibly their internal properties. The internal state of a particle recognizable in a comoving system of reference (rest system), is denoted here by z; let Z denote the set of internal states. For the purposes of collision theory the state of a free particle is determined by a pair $(z,V) \in Z{\times}V^3$, i.e. by an internal state z and a translation state V.

A (n,m)-collision is symbolized by

$$(z_\alpha, V_\alpha) \to (z_\beta, V_\beta), \tag{7}$$

independently of a frame of reference. On the left-hand-side the states of the n incident particles are indicated, on the right the m outgoing states are listed. A (2,1)-collision is also called a capture, a (2,2)-collision will be referred to as a binary collision.

With respect to an inertial system of reference S the world-velocities V_α determine 3-velocities $\vec{v}_\alpha$. We therefore also write, instead of (7),

$$S : (z,\vec{v}_\alpha) \to (z_\beta, \vec{v}_\beta), \tag{8}$$

and put $|\vec{v}_\alpha| = v_\alpha$ etc. Then $0 \le v_\alpha < \varepsilon^{-\frac{1}{2}}$, where $\varepsilon^{-\frac{1}{2}} = \infty$ if $\varepsilon=0$ by definition.

Two internal states z, z' are called collision-related if there exists either a capture

$$(z,V) + (z',V') \to (z'',V'') \tag{9}$$

with linearly independent world velocities V, V' or a binary collision

$$(z,V) + (z',V') \to (z'',V'') + (z''',V''') \tag{10}$$

such that there are three linearly independent vectors among
(V,V',V'',V''').

A function q : Z → R on the set of internal states is
called a <u>charge</u> if a (scalar) additive conservation law

$$\sum_{\alpha} q(z_{\alpha}) = \sum_{\beta} q(z_{\beta}) \tag{11}$$

holds for all those collisions which are possible according
to a collision theory. (According to present elementary particle theory there are at least four such charges: Q, electric
charge; B, baryon number; L_e, electron-lepton number and L_μ,
myon-lepton number; mass is not a charge. One may consider the
sum of the charges of one particular kind, of all the particles contained in some piece of material, as a measure of the
"amount of matter", or "quantum of substance" in the older
terminology.)

A collision theory formulates conditions which are intended to characterize all realizable collisions. In this
paper only conservation laws will be considered. They will
be assumed to have the same forms in all inertial frames of
reference.

Because of the principle of relativity the laws of conservation of the basic dynamical variables inertial mass,
linear momentum and energy are not independent of each other,
and the dependence of these variables on internal state (or
particle type) and speed is largely determined by that
principle and the group G_ϵ. Among the pioneers of collision
theory - J. Wallis, C. Wren and Ch. Huyghens - the last
author already made use of some of these dependences (see,
e.g. Mach 1883, chapter III, section 4). J.R. Schütz (1897)
realized that conservation of linear momentum is implied by
(Galilean) relativity and conservation of energy, provided
the quadratic dependence of energy on speed is taken for
granted. Special-relativistic collision theory was based on
conservation of linear momentum by G.N. Lewis and R.C. Tolman
1909 (see also R.C. Tolman 1912) and by H. Weyl 1923 who
emphasized that energy-conservation then follows, with $E=mc^2$.
P. Langevin developed relativistic collision theory solely
on the basis of energy conservation. (For original sources,
see R. Penrose and W. Rindler 1965.) The details of his
arguments have apparently not been published. Stimulated by
remarks about Langevin's work R. Penrose and W. Rindler gave
an argument along these lines in 1965 which was improved
shortly thereafter (J. Ehlers, R. Penrose and W. Rindler
(1965), see also W. Rindler (1966). These authors do not

assume, but derive the dependence of energy on internal state and speed. The presentation given below, while following essentially the last-mentioned papers, gives a more unified treatment of the Galilean and Einsteinian cases and exhibits the common transformation properties of the basic dynamical variables with respect to changes of the inertial coordinate systems.

As a basis of collision theory I take the following axioms:

(C_1) There exists an <u>energy function</u>

$$E : Z \times [0, \varepsilon^{-\frac{1}{2}}) \to R$$

such that for all collisions and all inertial frames of reference energy is additively conserved,

$$\sum_\alpha E(z_\alpha, v_\alpha) = \sum_\beta E(z_\beta, v_\beta). \tag{12}$$

(C_2) For each internal state z the function $E(z, \cdot)$ is continuous and non-constant on $[0, \varepsilon^{-\frac{1}{2}})$.

(C_3) Given an internal state z and a speed v, there exist internal states z', z" such that for any inertial frame S and any spatial unit vector $\vec{e}$,

$$S : (z, \vec{v}) + (z, -\vec{v}) + (z', \vec{0}) \to (z'', \vec{0}) \tag{13}$$

with $\vec{v} = v\vec{e}$ is a collision.

(C_4) Given two arbitrary internal states z, z', there exists a finite sequence $z = z_1, z_2, \ldots, z_n = z'$ of internal states such that any two consecutive members of it are collision related.

<u>Explanatory remarks.</u> In (C_1) it is essential that the conservation law (12) holds in <u>all</u> inertial frames with one universal, frame-independent function E. No transformation law for the values of E is postulated. – (C_2) serves to distinguish energy from charges. – Eq. (13) can be interpreted as the absorption of the particles with states $(z, \pm\vec{v})$ by a calorimeter which is and remains at rest and changes its internal state from z' to z". This process can be used to measure energy. (C_3) postulates that this kind of measurement is always possible. – (C_4) ensures that there are sufficiently many collisions in order that masses can be determined from collisions uniquely once a unit has been chosen (see below).

The following statements follow from the collision axioms:

<u>Proposition 1.</u> An energy function E uniquely determines two

functions m, U from Z to R such that for all states

$$E(z,v) = m(z)T_{\epsilon}(v) + U(z). \tag{14}$$

The range of m does not contain 0.
Proposition 2. If E is an energy function and m is defined by (14), then the world-vector conservation law

$$\sum_{\alpha} m(z_{\alpha})V_{\alpha} = \sum_{\beta} m(z_{\beta})V_{\beta} \tag{15}$$

is satisfied for all collisions.
Proposition 3. If E, E' are two energy functions and m,U; m',U' are the corresponding functions characterized by (14), then there is a unique real number e ≠ 0 such that

$$m' = em. \tag{16}$$

The function

$$E' - e E = U' - e U \tag{17}$$

is a charge.

These propositions can be proved as follows. Choose a fixed internal state z and speed v, and consider two collisions of the type (13), one along the x-axis, the other along the y-axis of S (see Fig. 1); that is possible according to (C3). View these collisions from an inertial frame S' which moves relative to S with velocity ǔ along the x-axis.

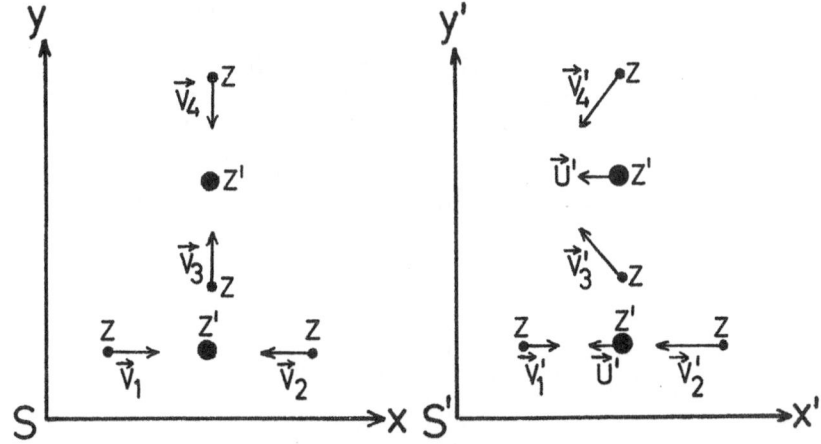

Fig. 1 (Qualitative) diagram showing the initial states of the particles involved in the two collisions of type (13) considered in the text, viewed from the centre-of-mass frame S and another frame S'.

Then, according to (C_1),

$$E(z,v_1') + E(z,v_2') + E(z',u') = E(z'',u')$$

and

$$E(z,v_3') + E(z,v_4') + E(z',u') = E(z'',u')$$

whence

$$E(z,v_3') + E(z,v_4') = E(z,v_1') + E(z,v_2'). \tag{18}$$

Next, use the fact that the function $v \to T_\varepsilon(v)$ defined in (5) is strictly monotonic and continuous. Therefore, we may temporarily use the variable $t=T_\varepsilon(v)$ instead of v and write, for fixed z, $\overline{E}(t)=E(z,v)$. Then the kinematical lemma - see eq. (6) - and (18) give the elementary functional equation

$$\overline{E}(\tfrac{1}{2}(t_1+t_2)) = \tfrac{1}{2}(\overline{E}(t_1) + \overline{E}(t_2)) \tag{19}$$

in which t_1 and t_2 vary independently since v and u can be so varied. (18) characterizes linear functions among continuous ones (see, e.g., Penrose and Rindler 1965), so $\overline{E}=mt+U$. Reintroducing z and v one obtains (14). Uniqueness of m and U follows from (14) and monotonicity of T_ε. The inequality $m(z) \neq 0$ follows from (14) and (C_2). This finishes the proof of proposition 1.

Proposition 2 is established separately for the cases $\varepsilon=0$ and $\varepsilon>0$. If $\varepsilon=0$ eq. (14) reads $E=\tfrac{1}{2}mv^2+U$. For an arbitrary collision (7), viewed from two inertial frames with relative velocity $\vec{u}$, energy conservation gives ($m_\alpha=m(z_\alpha)$ etc.)

$$\sum_\alpha (\tfrac{1}{2}m_\alpha \vec{v}_\alpha^2 + U_\alpha) = \sum_\beta (\tfrac{1}{2}m_\beta \vec{v}_\beta^2 + U_\beta)$$

and

$$\sum_\alpha (\tfrac{1}{2}m_\alpha (\vec{v}_\alpha+\vec{u})^2 + U_\alpha) = \sum_\beta (\tfrac{1}{2}m_\beta (\vec{v}_\beta+\vec{u})^2 U_\beta)$$

whence

$$\tfrac{1}{2}\vec{u}^2 (\sum_\alpha m_\alpha - \sum_\beta m_\beta) + \vec{u} :(\sum_\alpha m_\alpha \vec{v}_\alpha - \sum_\beta m_\beta \vec{v}_\beta) = 0.$$

Since $\vec{u}$ is arbitrary, this equation implies conservation of momentum, which because of $V=(1,\vec{v})$ proves proposition 2. If $\varepsilon>0$ we rewrite (14), using the definition (5), as $E=\varepsilon^{-1}m\gamma_\varepsilon+q$ where $q=U-\varepsilon^{-1}m$. Let W be the future-directed, timelike unit 4-vector directed along the time-axis of an inertial frame of reference S. Then $g(W,V)=W\cdot V=\gamma_\varepsilon(v)$ where v is the speed associated with V relative to S. The conservation of energy for the collision (7) with respect to S can thus be expressed as

$$\varepsilon^{-1}W\cdot(\sum_\alpha m_\alpha V_\alpha - \sum_\beta m_\beta V_\beta)+ (\sum_\alpha q_\alpha - \sum_\beta q_\beta) = 0. \qquad (20)$$

This equation must hold, for a fixed collision (7), for all W contained in V^3 (defined in section 2). Since V^3 is not contained in a hyperplane of V^4, the two expressions enclosed in round brackets in eq. (20) both vanish. This proves proposition 2 if $\varepsilon>0$. Moreover, it shows that the quantity

$$q := U-\varepsilon^{-1}m \qquad (21)$$

defined in terms of an energy function via proposition 1, is a charge as defined above.

To establish the third proposition one first notes that if two internal states z,z' are collision-related, the ratio $m(z')/m(z)$ can be determined uniquely from collision experiments by means of the conservation law (15). Assumption (C_4) ensures that this statement holds for arbitrary states z,z'. Having established the first claim of proposition 3 expressed by (16) one obtains (17) as an immediate consequence of (14) and (16). Since $E'-eE$ is conserved in all collisions so is $U'-eU$, but since this last function depends on z only it is a charge. This finishes the proof of proposition 3.

In view of propositions 1 to 3 it is reasonable to introduce the following terminology:
$m(z)$ is called the proper mass, $U(z)$ the internal energy (or rest energy) of the internal state z;
$m(z)T_\varepsilon(v)$ is called the kinetic energy, $m(z)\gamma_\varepsilon(v)$ the inertial mass and $m(z)\gamma_\varepsilon(v)\vec{v}$ the (linear) momentum of the state (z,V) relative to the respective inertial frame;
and $m(z)V$ is called the 4-momentum or mass-momentum world vector of the state (z,V).

Thus the inertial mass is defined as the factor of proportionality between velocity and momentum. According to proposition 2 the latter is conserved as is the inertial mass since the time- and space components of (15) give

$$\sum_{\alpha} m_{\alpha} \gamma_{\varepsilon\alpha} = \sum_{\beta} m_{\beta} \gamma_{\varepsilon\beta} \qquad\qquad (22)$$

and

$$\sum_{\alpha} m_{\alpha} \gamma_{\varepsilon\alpha} \vec{v}_{\alpha} = \sum_{\beta} m_{\beta} \gamma_{\varepsilon\beta} \vec{v}_{\beta}, \qquad\qquad (23)$$

respectively.

According to proposition 3 the collision laws (C_1)-(C_4) determine uniquely the proper and inertial masses, the kinetic energies, the momenta and 4-momenta of all states except for a common non-vanishing numerical factor, i.e. except for a mass-unit. For this reason "mass" is considered as a "physical dimension" independent of length and time, and the values of these dynamical variables are considered not as numbers, but as "physical quantities", i.e. elements of one-dimensional real vector spaces corresponding to the usual "physical dimensions".

An analogous uniqueness theorem for internal and total energies, U and E, is not implied by (C_1)-(C_4); for if E is an energy function and q is a charge, then $E'=E+q$ is an energy function, too. Thus even once a mass unit, and hence an energy unit has been selected, internal and total energies are fixed only up to a charge-function with the dimension of an energy. This lack of uniqueness can be removed in Einsteinian theory by a "natural" convention (see section 4) which fixes the zero-points of U and thus E, and leads to $E=m_{in}c^2$.

The empirical fact that all mass ratios are positive (for "massless" particles) is not implied by (C_1)-(C_4) but has to be introduced as an additional hypothesis, if needed.

The conserved quantities inertial mass, momentum and energy may be considered as components of a 5-vector ($m\gamma$, $m\gamma\vec{v}$,E) which transforms under changes of the frame of reference, according to a 5-dimensional linear representation of the isotropy group G_{ε}^{o} of spacetime. This 5-vector and its transformation law, and not the mass-momentum 4-vector, should be considered as the fundamental dynamical variable of collision theory, since it is this quantity which is defined for $\varepsilon=0$ as well as for $\varepsilon>0$ and the transformation law of which behaves continuously if ε tends to zero, as will be shown in the last section.

4. DIFFERENCES BETWEEN THE EINSTEINIAN AND GALILEAN
COLLISION THEORIES. THE FORMAL LIMIT $\varepsilon \to 0$ AND THE
APPROXIMATION RELATION.

The following theorems exhibit differences between the
two collision theories which are due to differences of the
underlying spacetime groups.
Proposition 4. If $\varepsilon = c^{-2} > 0$, any energy function can be decomposed uniquely as

$$E = mc^2 \gamma_\varepsilon + q .\tag{24}$$

The first summand is the only energy function which is the
time component of a (conserved) 4-vector; the second summand
is a charge.
Corollary. If $\varepsilon = c^{-2} > 0$, there exists exactly one energy
function which is a component of a quantity transforming
according to an irreducible representation of the homogeneous
Lorentz group G^0_ε. This energy function E obeys

$$E = m_{in} c^2 ,\tag{25}$$

and the associated internal energy U obeys

$$U = mc^2 .\tag{26}$$

Proposition 5. If $\varepsilon = 0$, proper and inertial masses are equal
and form a charge.
These propositions follow straightforwardly from the
theorems and definitions of section 3.
Proposition 4 and its corrollary show how it is possible, in the Einsteinian case, to remove the ambiguity in the
definition of energy discussed at the end of the preceding
section: One picks out that energy function which is pro-
portional to the time component of the 4-momentum. This
choice is in accordance with the general principle to define the basic physical variables such that they transform
according to indecomposable representations of the relativity group of the theory in question. (A representation is
said to be indecomposable (unzerfällbar) if it is not equivalent to a direct sum of two representations. A representation is called irreducible if its representation space does
not contain a non-trivial, invariant subspace. Irreducibility implies indecomposability. The converse is not true: The
representation

$$s \to \begin{pmatrix} 1 & s \\ 0 & 1 \end{pmatrix}$$

of the additive group of reals is indecomposable and reducible. See, e.g. H. Boerner (1967).)

It is clear from the preceding considerations that in the case $\varepsilon > 0$ the basic frame-independent, conserved quantities are the charges and the 4-momentum. They belong to irreducible representations of G_ε^o, the unit representation and the vector representation. The conservation law (15) with $m \neq 0$ can, in fact, replace the collision laws (C_1) and (C_2).

In the Galilean case the situation is different. Besides charges and 4-momentum, which again are indecomposable conserved quantities (though 4-momentum is reducible, containing the scalar mass as a constituent), there is the energy, linearly independent of them. To recognize its status from the point of view of the representations of the homogeneous Galilean group and, at the same time, to see how it arises from the Einsteinian theory in the limit $\varepsilon \to 0$, we consider, for $\varepsilon > 0$, the five-component quantity

$$P \oplus q \qquad\qquad\qquad (27)$$

where $P = mV$ is 4-momentum and q is a charge (with the dimension "energy"). The sign $\oplus$ indicates a direct sum; the expression (27) belongs to the (vector $\oplus$ scalar) representation of the homogeneous Lorentz group. Instead of using the components $(m\gamma_\varepsilon, m\gamma_\varepsilon \vec{v}, q)$ one can use the equivalent ones $(m\gamma_\varepsilon, m\gamma_\varepsilon \vec{v}, q + \varepsilon^{-1} m\gamma_\varepsilon) = (P, E)$ suggested by (24).

With respect to the special Lorentz transformation

$$t' = \gamma_\varepsilon(u)(t - \varepsilon u x_1), \quad x_1' = \gamma_\varepsilon(u)(x_1 - ut),$$
$$x_2' = x_2, \quad x_3' = x_3 \qquad\qquad (28)$$

(boost along the x_1-direction) the variables (P, E) transform according to the matrix

$$\begin{pmatrix} \gamma_\varepsilon(u) & , & -\varepsilon u \gamma_\varepsilon(u), & 0, & 0, & 0 \\ -u\gamma_\varepsilon(u), & & \gamma_\varepsilon(u) & , 1, & 0, & 0 \\ 0 & , & 0 & , 1, & 0, & 0 \\ 0 & , & 0 & , 0, & 1, & 0 \\ T_\varepsilon(u) & , & -u\gamma_\varepsilon(u) & , 0, & 0, & 1 \end{pmatrix} \qquad (29)$$

For fixed speed u this matrix depends continuously on ε even at $\varepsilon=0$; its value at $\varepsilon=0$ is

$$
\begin{pmatrix}
1 & 0 & 0 & 0 & 0 \\
-u & & & & 0 \\
0 & & 1_3 & & 0 \\
0 & & & & 0 \\
\tfrac{1}{2}u^2 & -u & 0 & 0 & 1
\end{pmatrix}
\tag{30}
$$

where 1_3 denotes the 3×3 unit matrix. Thus in the limit the "5-vector" $(m,m\vec{v},E)$ of mass, momentum and energy transforms according to the indecomposable representation given by (30). This representation arises by "contraction" $\varepsilon \to 0$ from the (directly decomposable!) vector $\oplus$ scalar representation of the Lorentz group; it contains as constituents the 4-vector representation and the scalar representation corresponding to P and m, respectively. This clarifies the peculiar status of the energy with respect to Galilean transformations. (Note that q in (27) denotes any charge, e.g. q=0.)

The exposition given so far exhibits the common structure as well as the differences between the collision theories based on Galilean and Einsteinian kinematics, respectively. It shows how laws of the latter theory go over into those of the former one under the formal limit $\varepsilon \to 0$. It also shows that the concepts of mass, energy, momentum etc. in both theories have the same "meaning" insofar as they are characterized by analogous relations to directly measureable velocities.

These relations also indicate the conditions under which quantitative statements of the Galilean theory are approximately correct in Einstein's theory. To obtain error estimates it suffices to determine upper bounds for the differences between the values of relevant functions at the "true" value $\varepsilon = c^{-2}$ and the fictitious value $\varepsilon = 0$. For example, if speeds are restricted to the range $0 \le \frac{v}{c} \le \frac{1}{10}$, then the inequalities

$$
\left| \gamma_\varepsilon(v) - (1 + \tfrac{1}{2}(\tfrac{v}{c})^2) \right| \le \tfrac{1}{2}(\tfrac{v}{c})^4 \le 5 \cdot 10^{-5}
$$

and $\quad \left| m\gamma_\varepsilon(v)\vec{v} - m\vec{v} \right| \le (\tfrac{v}{c})^2 mv \le 10^{-2} \, mv$

and $\quad \left| mc^2\gamma_\varepsilon(v) - (mc^2 + \tfrac{1}{2}mv^2) \right| \le \tfrac{1}{2}mv^2(\tfrac{v}{c})^2 \le 10^{-2} \, \tfrac{1}{2}mv^2$

show the relative accuracy with which the Einsteinian expressions for momentum and energy may be replaced by the
Galilean ones. If one chooses a smaller bound for $\frac{v}{c}$ than 0.1
the bounds for the relative errors decrease. The point in
mentioning these well known elementary facts is only to
illustrate the usefulness of having a common, parameter-
dependent formulation of the old and new theories in order
to obtain error estimates for approximations.

As an example of the frequently encountered circumstance
that an empirical law appears in the old theory as an exact
consequence of the basic laws whereas in the new theory it
figures as approximately valid only, we may consider the
law of conservation of mass in chemical reactions, experimentally established by Lommonossow, Lavoisier, Landolt and
others. In the old collision theory, combined with the kinetic theory of matter and heat, this empirical law follows
trivially from the law of conservation of mass. In the new
theory this explanation does not apply to isothermal reactions, for the heat w carried away or supplied changes
the inertial mass by w/c^2. Only if the molecular structure
of matter, the smallness of the binding energies of the
molecules in comparison with the rest energies of their constituent nuclei and electrons, and the conservation of the
nuclei and electrons in chemical reactions are taken into
account, the law of conservation of mass for isothermal reactions follows as an approximation. This explanation is more
complicated than the older one. On the other hand the Ein-
stein-based collision theory predicts measurable deviations
from conservation of mass which are known to be true, an
example showing that the new theory is an improvement on the
old one.

At least in the case considered here, but presumably
also in more complicated cases, one can clearly argue for a
continuity of meaning of concepts and for progress in the
succession of theories. From the point of view of the new
theory one can understand
1) in which domain the old and the new theory are nearly
   identical - here for v<<c;
2) in which domain the new theory represents a significant
   improvement in precision in comparison to the old one -
   here for v≲c;
3) in which domain the old theory is fictitious, i.e. does
   not correspond to real phenomena - here Galilean collision theory for particle speeds exceeding the speed of
   light.

Jürgen Ehlers
Max-Planck-Institut für Physik und Astrophysik
Institut für Astrophysik
Karl-Schwarzschild-Str. 1
D-8046 Garching b. München

REFERENCES

A. Ashtekar, Commun. Math. Phys. 71, 59 (1980)

H. Boerner, Darstellungen von Gruppen, Springer, 2. Aufl. Berlin 1967

E. Cartan, Ann. Éc. Norm. 41, 1 (1924).

R.A. Coleman and H. Korte, A Realist Field Ontology of the Causal-Inertial Structure. Preprint, Department of Philosophy, University of Regina, Canada, 1981

G. Dautcourt, Acta Physica Polonica 25, 637 (1964).

J. Ehlers, R. Penrose and W. Rindler, Am. J. Phys. 33, 995 (1965)(1965)

J. Ehlers, Über den Newtonschen Grenzwert der Einsteinschen Gravitationstheorie. To appear in Festschrift für P. Mittelstaedt; J. Nitsch, J. Pfarr and E.W. Stachov (eds.).

A. Einstein, Annalen der Physik 17, 891 (1905).

K. Friedrichs, Math. Annalen 98, 566 (1927).

K. Hepp, Commun. Math. Phys. 35, 265 (1974).

W. Kundt, Springer Tracts in Modern Physics 40, 106 (1966).

H.P. Künzle, Ann. Inst. Henri Poincaré 17, 337 (1972).

H.P. Künzle, General Relativity and Gravitation 7, 445 (1976).

G.N. Lewis and R.C. Tolman, Phil. Mag. 18, 510 (1909); R.C. Tolman, Phil. Mag. 23, 375 (1912).

E. Mach, Die Mechanik, Historisch-Kritisch Dargestellt, Erstauflage Leipzig 1883. Photomech. reproduction of the 9th edition (1933) by Wissenschaftl. Buchgesellschaft Darmstadt 1973.

R. Penrose and W. Rindler, Am. J. Phys. 33, 55 (1965).

W. Rindler, Special Relativity, 2. Aufl., Intersc. Publ., Inc., John Wiley And Sons, New York 1966, esp. ch. V.

W. Rindler, Essential Relativity, Special, General, and Cosmological, revised 2nd ed., Springer, New York 1979, in part. section 2.17 and 2.6, footnote 8, p. 31.

E. Scheibe, Eine Fallstudie zur Grenzfallbeziehung in der Quantenmechanik. To appear in Festschrift für P. Mittelstaedt; J. Nitsch, J. Pfarr and E.W. Stachov (eds.).

J.R. Schütz, Göttinger Nachrichten S. 110, 1897.

H. Weyl, Raum, Zeit, Materie, 5. Aufl., Springer, Berlin 1923.

C.F. von Weizsäcker

GEOMETRIE UND PHYSIK

Behauptung: Die Geometrie ist ein Teil der Physik. Frage:
Was ist der Sinn dieser Behauptung? Vermutung: Es gibt ein'
System geometrischer Axiome, die dasjenige korrekt beschreiben, was in der Physik der Raum genannt wird. Rückfrage: Wie entscheidet man, ob diese Beschreibung korrekt
ist? Weitere Vermutung: durch Erfahrung. Erneute Rückfrage:
Kann man durch Erfahrung über geometrische Sätze entscheiden? Gegenfrage: Wie entscheidet man denn über irgend-
welche physikalischen Sätze durch Erfahrung?
    Der vorliegende Aufsatz präsentiert eine erkenntnis-
theoretische (Abschnitt 7) und eine physikalische (Abschnitt 10) Hypothese über diesen Fragenkreis. Eine zu-
sammenfassende Reflexion enthält Abschnitt 11.

1. AXIOMATIK

In der älteren Tradition der abendländischen Wissenschaft
erscheinen Geometrie und Physik als streng getrennte
Disziplinen. Geometrie ist in dieser Auffassung ein Teil
der Mathematik, deren Erkenntnisse von der Erfahrung unabhängig sind, Physik hingegen eine Wissenschaft, deren
fundamentale Sätze durch Erfahrung bewiesen werden müssen
und können. Wir sind heute skeptisch genug,keine dieser
Meinungen für selbstverständlich zu halten. Eben dadurch
sind wir dafür vorbereitet, die Entstehung dieser Meinungen als wissenschaftsgeschichtlichen Vorgang zu würdigen.
    Die Babylonier besassen den Inhalt des Lehrsatzes des
Pythagoras tausend Jahre vor Pythagoras. Man kann vermuten,
dass das Selbstverständnis des geometrischen Wissens, das
die Griechen vom Orient übernahmen, durch den Wortsinn des
griechischen Wortes geo-metria angedeutet ist: Erdvermes-
sung. Es scheint die Entdeckung der griechischen Mathematiker gewesen zu sein, dass die geometrischen Sachverhalte
in Lehrsätzen formuliert werden können, die sich logisch
aus einer kleinen Zahl von Ausgangssätzen herleiten lassen.
Mit einer Vergröberung griechischer Distinktionen nennen
wir heute alle diese Ausgangssätze Axiome. Wenn man von

D. Mayr and G. Süssmann (eds.), Space, Time, and Mechanics, 39–86.
Copyright © 1983 *by D. Reidel Publishing Company.*

den immanenten Problemen der Logik absieht, reduziert sich
die Frage nach der Wahrheit der Geometrie dann auf die
Frage nach der Wahrheit (oder dem Sinn) ihrer Axiome.

Die vorherrschende Meinung der älteren neuzeitlichen
Wissenschaft war, dass die geometrischen Axiome evident,
also unabhängig von der Erfahrung als wahr einleuchtend
sind, dass sie aber mit der Welt der Erfahrung den Zusammenhang haben, sich in ihr stets zu bewähren. Im Beispiel:
Man kann aus evidenten Axiomen logisch folgern, dass die
Winkelsumme im Dreieck gleich zwei Rechten ist; und wenn
man ein physisches Dreieck vermisst, so wird man (innerhalb der Fehlergrenzen der Messung) stets zwei Rechte
als seine Winkelsumme finden. Diese Beschreibung fordert
zu der erkenntnistheoretischen Frage heraus, wie denn
diese Harmonie zwischen Evidenz und Erfahrung garantiert
ist. Man darf wohl den allgemeinen Konsens der heutigen
Wissenschaftstheoretiker dafür voraussetzen, dass schon
diese Beschreibung selbst zu naiv ist, als dass sie eine
Beantwortung der Frage zuliesse; man muss die in ihr implizierten Auffassungen sowohl von Evidenz wie von Erfahrung zunächst auflösen.

Nur als historische Randbemerkung sei gesagt, dass
die Griechen an dieser Naivität unschuldig sind. Platon
wusste, dass geometrische Sätze in der Erfahrung niemals
in Strenge überprüft werden können, und dass dasselbe für
die Sätze der Physik gilt, d.h. er kritisierte die Naivität des Empirismus. Er wusste ferner, dass die Mathematiker vom Sinn ihrer Grundbegriffe nicht Rechenschaft
geben können, sondern dass sie diesen Sinn schlicht unter-
stellen (ihre Grundbegriffe sind "hypotheseis" = Unter-
Stellungen), d.h. er kritisierte die Naivität des Evidenzbegriffs. Platons eigener Versuch eines Aufbaus der mathematischen Physik im Timaios kann nur dann in seiner begrifflichen Struktur verstanden werden, wenn man sieht,
dass in ihm aus für Platon zwingenden systematischen
Gründen der Unterschied zwischen Mathematik und Physik von
vornherein gar nicht gemacht wird. Ferner hat Imre Toth
(1967) sehr starke Gründe für die These vorgebracht, dass
die Auffassung der Geometrie durch Aristoteles einen
Wissensstand der ihm zeitgenössischen Mathematiker voraussetzt, dem die Möglichkeit einer Axiomatisierung der
Geometrie ohne das "euklidische" Parallelenpostulat geläufig war. Das methodologische Niveau der griechischen
Wissenschaft und Philosophie war offenbar so viel höher
als das ihrer neuzeitlichen Nachfolger, dass wir bis in

unsere Tage haben warten müssen, um einige der Fragen
wieder zu entdecken, welche die Griechen mit ihren uns
überlieferten Lehren zu lösen versucht haben. Im folgenden
will ich aber die neuzeitlichen Probleme nur im neuzeitlichen Kontext behandeln.

In der Geschichte der Geometrie wurde die Naivität
des Evidenzbegriffs am Beispiel des euklidischen Paralle-
lenpostulats aufgelöst. Schon der Versuch, dieses Postulat
aus anderen Axiomen zu beweisen, zeigt, dass man es faktisch nicht als evident empfand. Die Unmöglichkeit dieses
Beweises wurde durch den positiven Aufbau einer nichteuklidischen Geometrie demonstriert. Dieser Nachweis verlangt
aber, um streng zu sein, den Beweis der Widerspruchsfreiheit der nichteuklidischen Geometrie. Dieser wurde zuerst
erbracht durch die Konstruktion euklidischer Modelle nichteuklidischer Räume (populärstes Beispiel: die Kugelfläche
als Modell der "Ebene" der sphärischen Geometrie). Methodisch wichtig ist hieran u.a. die "konventionalistische"
Verwendung der Vokabeln der Geometrie wie "Gerade" und
"Ebene", um Gegenstände zu bezeichnen, die nach dem
"evidenten" anschaulichen Verständnis der bisherigen
Geometrie die Eigenschaften, "gerade" bzw. "eben" zu sein,
gar nicht haben. Thematisiert wird diese Methode in der von
Hilbert eingeführten und heute unter Mathematikern herr-
schenden Auffassung von Axiomatik. Nach ihr ist weder die
Wahrheit der Axiome noch auch nur die Bedeutung der in
ihnen verwendeten Worte ein Thema der Mathematik (Hilbert:
Statt "Punkt", "Gerade", "Ebene" hätte ich genau so gut
sagen können "Liebe", "Gesetz", "Schornsteinfeger"; die
axiomatische Mathematik in diesem Sinne des Wortes befasst
sich nur mit den logischen Beziehungen zwischen formal
präzisierten Sätzen.

Die mit der Gabel der Axiomatik ausgetriebenen inhaltlichen Probleme der Mathematik kehren bekanntlich durch das
Fenster der Meta-Mathematik zurück (Horaz, Epistel I, 10,
24). Die Widerspruchsfreiheit der Geometrie wird durch
Reduktion auf die Arithmetik bewiesen. Die Widerspruchsfreiheit der Arithmetik verlangt zu ihrem Nachweis inhaltliche metamathematische Überlegungen, welche de facto ein
Stück geometrischer Anschauung von Zeichenreihen etc. enthalten dürften. Doch ist auch die Metamathematik nicht
Gegenstand dieses Aufsatzes. Ich will mich im folgenden
auf den Standpunkt stellen, dass uns die mathematische
Analyse der Grundlagen der Geometrie eine beliebige Menge
möglicher geometrischer Axiomensysteme zur Verfügung stellt.

Die Frage ist dann, ob eines von ihnen, wenn ja welches,
oder ob vielleicht mehrere von ihnen das zu beschreiben
geeignet sind, was wir in der Physik den Raum nennen.

2. EMPIRISMUS

Gauss hat im Zug der Hannoverschen Landesvermessung das
grosse Dreieck zwischen den Bergen Brocken-Inselberg-Hoher
Hagen vermessen. Er wusste, dass in der nichteuklidischen
Geometrie die Winkelsumme eines Dreiecks um so weiter von
zwei Rechten abweicht, je grösser das Dreieck ist. Er
registrierte, dass er innerhalb der Fehlergrenzen keine
Abweichung fand. Er hat also die Möglichkeit in Betracht
gezogen, dass in der Wirklichkeit eine nichteuklidische
Geometrie gelten könnte, und dass hierüber empirisch ent-
schieden werden könnte. Diese Denkmöglichkeit war den
Mathematikern des späteren 19. Jahrhunderts geläufig.
Einstein hat auf Grund theoretisch-physikalischer Überlegungen die Hypothese der Geltung einer bestimmten, nämlich
der Riemannschen Geometrie in der Wirklichkeit aufgestellt.
Er hat versucht, die Konsequenzen der diese Hypothese ent-
haltenden Allgemeinen Relativitätstheorie (ARTh) einer
empirischen Prüfung zugänglich zu machen. Seit Einstein
glauben fast alle Physiker, dass über die in der Wirklichkeit geltende Geometrie eine empirische Entscheidung
möglich sei. Diese These sei im folgenden als Empirismus
(genauer: geometrischer Empirismus) bezeichnet.
     Dieser Empirismus ist nun in seiner schlichten Form
ebenso naiv wie der Glaube an die Evidenz der geometrischen Axiome. Ich werde im Abschnitt 7 eine präzisierte
Fassung des Empirismus vorschlagen, in der er m.E. aufrecht-
erhalten werden kann. Dazu ist es aber zunächst nötig,
seine naive Fassung aufzulösen. Wir wollen diese als
direkten geometrischen Empirismus bezeichnen, d.h. als den
Glauben an die direkte empirische Entscheidbarkeit von
Sätzen der physikalischen Geometrie.
     Nehmen wir an, Gauss oder irgend ein heutiger Beob-
achter hätte empirisch in einem Lichtstrahlendreieck eine
Abweichung der Winkelsumme von zwei Rechten gefunden und
diese Beobachtung sei als reproduzierbar anerkannt. Spä-
testens seit Poincaré wird nun gegen den Empirismus so
argumentiert: In diesem Falle hätte der Beobachter nicht
die Gültigkeit einer nichteuklidischen Geometrie empirisch
bewiesen. Wenigstens müsste er Gründe dafür angeben, dass
er die physikalisch näherliegende Deutung vermeidet, die

Lichtstrahlen seien keine geraden Linien. (Die übliche
Ausdrucksweise für die Lichtablenkung am Sonnenrand, die
als eine empirische Bestätigung der Relativitätstheorie
gilt, ist "Krümmung der Lichtstrahlen im Schwerefeld").
Dieser Einwand zeigt jedenfalls, dass der Empirist ohne
eine Theorie des gemessenen Vorgangs nichts beweisen kann.

Ich möchte zunächst zwei Thesen betrachten, die in der
Kritik am Empirismus weitergehen und sogar behaupten, man
könne durch rein erkenntnistheoretische Überlegung einsehen, dass man grundsätzlich keine empirische Entscheidung über die Geometrie tragen kann. Sie sollen hier als
Hierarchismus und als Konventionalismus bezeichnet werden.
Ich halte beide für falsch, glaube aber, dass vor allem
der Konventionalismus eine sehr starke Position ist, ohne
deren volles Verständnis unser Problem nicht gelöst werden
kann.

3. HIERARCHISMUS

Hierarchismus ist kein üblicher Terminus der Wissenschaftstheorie. Ich verwende dieses Wort, um eine Ansicht zu bezeichnen, die dort, wo man sie für wahr hält, gewöhnlich
als so sebstverständlich erscheint, dass man vergisst, sie
als besondere Voraussetzung auszusprechen. Es ist die Ansicht, dass eine eine Hierarchie der Wissenschaften gebe,
in der die jeweils niedrigeren die jeweils höheren zur
methodischen Voraussetzung haben, aber nicht umgekehrt.
Z.B. gilt die Logik als die hierarchisch höchste der
Wissenschaften: alle Wissenschaften haben die Logik zur
methodischen Voraussetzung, denn sie müssen gemäss den
Regeln der Logik verfahren; die Logik aber hat keine von
ihnen zur methodischen Voraussetzung, denn (so meint man)
logische Wahrheiten können nicht von den Ergebnissen von
Einzelwissenschaften abhängen. Die Leugnung dieses hierarchischen Verhältnisses zwischen der Logik und den anderen
Wissenschaften erweckt unmittelbar den Verdacht des cir-
culus vitiosus. Analog sieht man auch das Verhältnis
zwischen der Mathematik und den empirischen Wissenschaften:
zur Aufstellung und Prüfung empirischer Gesetze braucht
man Mathematik und schon darum erscheint es methodisch
unsauber, eine Abhängigkeit der inhaltlichen Wahrheit
der Mathematik von der Erfahrung anzunehmen.

Ich möchte die Vermutung aussprechen, dass der Hierarchismus grundsätzlich und in allen Fällen falsch ist, dass
vielmehr sowohl zwischen der Logik und den Wissenschaften

wie zwischen der Mathematik und den empirischen Wissenschaften ein Verhältnis gegenseitiger methodischer Abhängigkeit besteht. Diese allgemeine wissenschaftstheoretische These wird, soweit hier notwendig, im Abschnitt 7
besprochen werden. Im Augenblick sei nur zweierlei hervorgehoben: die historische Herkunft des Hierarchismus
und seine Bedeutung für die Geometrie.

Der Hierarchismus ist eine Folge der griechischen
Entdeckung der Möglichkeit einer axiomatischen Mathematik.
In einem fest vorgegebenen axiomatischen System werden die
Theoreme aus den Axiomen logisch hergeleitet. Die Wahrheit
der Theoreme hat also die Logik und die Wahrheit der Axiome zur (hinreichenden) Bedingung. Die Logik und die
Axiome müssen als evident vorausgesetzt oder von noch
höheren Voraussetzungen her begründet werden. Aristoteles
hat in den Analytica Posteriora den Begriff der deduktiven
Wissenschaft gemäss diesem Schema entworfen. Tatsächlich
hat es aber ausser der Mathematik und der mathematischen
Logik nie eine deduktive Wissenschaft gegeben. Gleichwohl
hat man in der europäischen Tradition sowohl die Philosophie wie die empirischen Wissenschaften an diesem Ideal
gemessen. Der Hierarchismus ist gleichsam der regulative
Gebrauch dieser Idee von Wissenschaft. Dabei hat man de
facto die hierarchische Überodnung der Logik und Mathematik über die empirischen Wissenschaften nicht im Sinne
einer Deduzierbarkeit dieser aus jener interpretiert,
sondern einer Unabhängigkeit jener von diesen. Das ist
aber nicht etwa eine schwächere, sondern eine stärkere
Behauptung. Ist B aus A deduzierbar, so ist eine Widerlegung von B zugleich eine Widerlegung von A. Der Hierarchismus übernimmt gerade nicht die einwandfreie logische
Struktur der deduktiven Wissenschaft, sondern ihre frag-
würdige Annahme evidenter Axiome.

Die Geometrie ist der Ort, an dem der Hierarchismus
zuerst erschüttert worden ist. Die Geometrie galt seit
den Griechen als Teil der Mathematik. Also musste sie der
Physik hierarchisch übergeordnet sein. Wenn zuerst Mathematiker und dann Physiker die empirische Entscheidung
über die Geltung gewisser geometrischer Axiome in der
Wirklichkeit für möglich hielten, so verletzten sie diese
Vorstellung. Es gab aber zwei Möglichkeiten, das Problem
der empirischen Geltung der Geometrie so zu beurteilen,
dass das Prinzip des Hierarchismus unangetastet blieb.

Die moderne (Hilbertsche) Auffassung der Axiomatik
ist der eine Ausweg, und zwar der weichere von den beiden.

Man schränkt den hierarchischen Anspruch der Mathematik in der Geometrie auf den logischen Zusammenhang zwischen Axiomen und Theoremen ein. Damit ist die eigentliche Substanz dessen, was man seit den Griechen unter Geometrie verstanden hat, nämlich der inhaltliche Sinn ihrer Begriffe und die Wahrheit ihrer Axiome, aus der Mathematik und damit aus dem Anspruch hierarchischer Überordnung ausgeschlossen. Man kann dies, im Gegensatz zum inhaltlichen Hierarchismus der älteren Auffassung als formalen Hierarchismus bezeichnen.

Nun ist aber auch der inhaltliche Hierarchismus nicht eine unsinnige Ansicht, sondern nur die dogmatische Verfestigung wirklicher Unsymmetrien zwischen den Wissenschaften; niemand wird z.B. die Arithmetik im selben Sinne für empirisch halten wie die Botanik. Ich möchte daher zunächst auch im Fall der Geometrie die Argumentationsstrategie verfolgen, dem inhaltlichen Hierarchismus so weit wie möglich entgegenzukommen. Hier bietet sich der zweite, harte Ausweg, nämlich das strikte Festhalten an der überlieferten Geometrie. Man sagt etwa: Die euklidische Geometrie ist a priori gewiss. Wenn in einem empirisch hergestellten Dreieck die Winkelsumme nicht den aus der euklidischen Geometrie folgenden Wert hat, so ist a priori gewiss, dass die Seiten dieses Dreiecks keine Geraden sind. Im Beispiel des Lichtstrahlendreiecks führt also gerade die Gewissheit der euklidischen Geometrie zu einer eindeutigen Folgerung für die Optik: die Lichtstrahlen, welche durch ein Schwerefeld gehen, sind keine Geraden.

Wenn der Empirist diese These widerlegen wollte, müsste er zeigen, dass es keine Interpretation der bekannten Erfahrung geben kann, die mit einer euklidischen Beschreibung der Phänomene verträglich wäre. Dieser Beweis lässt sich, wie alle Unmöglichkeitsbeweise im empirischen Bereich, voraussichtlich überhaupt nicht in Strenge führen. Der Empirist wird daher zum Gegenangriff übergehen und fragen, womit sein Gegner die Apriori-Gewissheit der euklidischen Geometrie begründen will, nachdem die logische Möglichkeit nichteuklidischer Geometrien bekannt geworden ist. In den nächsten drei Abschnitten verfolgen wir u.a. Argumentationen über dieses Problem.

4. KANTS FRAGESTELLUNG

Kants Auffassung der Geometrie ist insofern überholt, als sie vor der expliziten Aufstellung nichteuklidischer

Geometrien entworfen wurde. Andererseits bleibt sie auch
heute lehrreich, weil Kant zwar am inhaltlichen Hierarchismus streng festhielt, ihn aber nicht schlicht behaup-
tete, sondern die Notwendigkeit betonte, ihn detailliert
zu begründen. Hier sei nur soviel von seiner Auffassung
skizziert als wir im folgenden brauchen werden.

Kant kannte das Ergebnis von Saccheri und Lambert,
dass das Parallelenpostulat nicht aus den anderen Axiomen
der Geometrie logisch hergeleitet werden kann. In diesem
eingeschränkten Sinne war ihm die logische Möglichkeit
einer nichteuklidischen Geometrie vertraut. Damit stand
für ihn das Parallelenpostulat aber in einer Linie mit
allen anderen fundamentalen Einsichten der Mathematik.
Sie waren nicht aus höheren Prinzipien logisch herleit-
bar und gleichwohl a priori gewiss; sie waren, um seinen
Ausdruck zu gebrauchen, synthetische Urteile a priori.
Die Grundfrage seiner Erkenntnistheorie war daher: wie
sind synthetische Urteile a priori möglich?

Diese Auffassung steht in der Mitte zwischen zwei
bequemeren, aber nach Kants Überzeugung unhaltbaren Ansichten. Nach dem Logizismus sind die Urteile der Mathematik einschliesslich der Axiome a priori, aber analytisch; sie sind a priori, weil sie logisch notwendig
sind. Dies kann für die Axiome der Geometrie mit Sicherheit bestritten werden; Kant bestreitet es auch für die
Grundlagen der Arithmetik. Nach dem radikalen Empirismus
müssten umgekehrt die Urteile der Mathematik synthetisch,
aber a posteriori, also auf Erfahrung gegründet sein.
Für die inhaltlich gedeuteten Axiome der Geometrie ist
dies heute die herrschende Ansicht der Physiker. Für die
Arithmetik aber erscheint der radikale Empirismus schwer
durchführbar. Kant jedenfalls verwarf ihn für beide Zweige
der ihm bekannten Mathematik mit der Begründung, dass
Erfahrung die ihnen eignende Notwendigkeit und Gewissheit grundsätzlich nicht garantieren kann.

Soweit ist aber nur Kants Problem formuliert. Wie ist
Mathematik möglich, wenn sie weder analytisch noch a
posteriori ist? Kants Antwort beruht auf seiner Unterscheidung von Anschauung und Denken. Anschauung ist Rezep-
tivität, Denken ist Spontaneität des menschlichen (d.h.
endlichen) Bewusstseins. Synthetische Erkenntnis muss
auf Anschauung beruhen, denn sie fügt zu den Begriffen
etwas hinzu, was in ihnen nicht schon enthalten war. Erkenntnis a priori aber kann nicht auf der Erfahrung beruhen; das ist ihre Definition. Synthetische Erkenntnis

a priori muss also auf Anschauung beruhen, die nicht Erfahrung ist. Solche Anschauung nennt Kant reine Anschauung. Er findet sie in den Formen aller Anschauung, d.h. in der Zeit und im Raum. Mathematik beruht auf der Konstruktion der Begriffe in der reinen Anschauung, Arithmetik in der Zeit, Geometrie im Raum.

Wir können uns auf die Details dieser sehr voraussetzungsvollen Theorie hier nicht einlassen. Für die Arithmetik sei nur bemerkt, dass sie eng verwandt ist mit dem Intuitionismus Brouwers und dem Konstruktivismus, wie ihn z.B. Lorenzen vertritt. Sie ist also von aktuellem Interesse. Für die Geometrie freilich würden fast alle heutigen Mathematiker ihre Begründung auf Konstruktion in der reinen Anschauung Raum verwerfen (für Lorenzen vgl. jedoch Abschnitt 6). Gerade sie muss uns aber hier interessieren.

Kant beansprucht, mit seiner Theorie der Mathematik zugleich ein Problem zu lösen, das die Mathematiker meist nicht beschäftigt, das aber für die Erkenntnistheorie der Physik fundamental ist: das Problem der Geltung mathematischer Gesetze in der Erfahrung. Wer mit Hume erkannt hat, dass aus der empirischen Geltung von Gesetzen in der Vergangenheit ihre allgemeine Geltung und damit ihre Geltung in der Zukunft logisch schlechterdings nicht abgeleitet werden kann, der steht vor diesem Problem. Es scheint, dass die moderne empiristische Wissenschaftstheorie erst jetzt zu realisieren beginnt, dass sie dieses Problem nie gelöst hat und grundsätzlich nicht lösen kann. Kants Lösungsvorschlag ist: Die reine Anschauung (Zeit und Raum) ist zugleich die Form aller empirischen Anschauung. Deshalb müssen Sätze, die durch Konstruktion in der reinen Anschauung begründet sind, in jeder empirischen Anschauung gelten. Ich übergehe wieder die sehr schwierige Frage, was die Gleichsetzung von reiner Anschauung und Form aller Anschauung bedeuten soll und hebe nur das erkenntnistheoretische Ziel dieses Lösungsvorschlags hervor. Er ist ein Spezialfall der allgemeinen These Kants zur Lösung des Humeschen Problems. Sätze, die in jeder Erfahrung gelten sollen, können sich weder durch spezielle Erfahrung begründen lassen, noch können sie eine Begründung haben, die mit Erfahrung gar nichts zu tun hat. Sie müssen vielmehr aus den Bedingungen jeder möglichen Erfahrung folgen. Dann haben sie einen Bezug auf jede mögliche Erfahrung und man kann doch a priori wissen, dass sie nicht durch Erfahrung widerlegt werden können. Von dieser These werde ich

(Abschnitt 7) die Vermutung übernehmen, dass allgemein empirisch gültige Gesetze Bedingungen aller Erfahrung formulieren, aber nicht ihre hierarchistische Verengung, dass
unsere Formulierungen solcher Gesetze nicht durch Erfahrung
korrigiert werden könnten.

Eine Anwendung dieser Gedanken auf die physikalische
Geometrie liegt in der Messtheorie. Bohrs These, dass ein
physisches Gebilde nur als Messapparat geeignet ist, wenn
wir es in Raum und Zeit der Anschauung kausal beschreiben
können, ist gut kantisch. Hier stellt sich aber die Frage,
ob der Raum unserer Anschauung denn euklidisch ist. Die
Antwort der empirischen Psychologie muss wahrscheinlich
lauten, dass der Raum unseres Anschauungsvermögens und
unserer Phantasie weder euklidisch noch nichteuklidisch,
sondern unpräzise ist. In der anschaulichen Vorstellung
können wir zwischen einem Tausendeck und einem Zehntausend-
eck, zwischen einer Million und einer Milliarde Kilometern,
zwischen 10^{-8} und 10^{-12}cm nicht unterscheiden. Ein Kanti-
aner würde vielleicht einwenden, dies gelte zwar für die
empirische, nicht aber für die reine Anschauung. Dieser
Einwand fordert aber die Frage heraus, in welchem Sinne es
eine reine Anschauung gibt. In den metaphysischen Anfangsgründen der Naturwissenschaft nennt Kant den Raum eine Idee.
Dort geht es um die Absolutheit des Raums, die nicht der
Anschauung, sondern der Vernunft zugehört. Für unser Problem wird man sagen: die scheinbar anschauliche Evidenz der
präzisierten Geometrie beruht darauf, dass wir uns erlauben,
extreme Grössen ähnlich verkleinert oder vergrössert vorzu-
stellen. Die Existenz ähnlicher Figuren ist aber bereits
ein dem Parallelenpostulat äquivalentes Postulat der euklidischen Geometrie. Die reine Anschauung Kants ist also
selbst schon ein Produkt des Denkens (vgl. Kritik der reinen Vernunft, 2. Auflage, S. 161 Fussnote).

Wir werden keine der speziellen Thesen Kants übernehmen, sondern nur seine Fragestellung, ob Geometrie Bedingungen der Möglichkeit von Erfahrung formuliere.

5. KONVENTIONALISMUS

Ehe wir den letzten Versuch besprechen, die A priori-
Gewissheit der euklidischen Geometrie zu begründen, nämlich
den Versuch von Dingler und Lorenzen, müssen wir den von
ihm methodisch vorausgesetzten Konventionalismus erörtern.

Betrachten wir als Beispiel noch einmal den fiktiven
Fall, Gauss hätte im grossen optischen Dreieck eine Winkel-

summe ungleich zwei Rechten gefunden. Der geometrische
Empirismus in naiver Fassung hätte gefolgert, in der Natur
gelte eine nichteuklidische Geometrie; der Hierarchismus
hätte gefolgert, Lichtstrahlen seien keine Geraden. Der
Konventionalismus würde sagen, beide Beschreibungsweisen
seien zulässig. Es sei ein Missverständnis des konventionellen Charakters unserer Sprache, eine von beiden als die
empirisch richtige auszeichnen zu wollen.

Ein zweites Beispiel: Vor etwa 40 und noch einmal vor
etwa 20 Jahren konnte man in Bahnhofskiosken Schriften über
die sogenannte Hohlwelttheorie kaufen. Nach dieser Theorie
ist die Erde eine Hohlkugel, auf deren innerer Oberfläche
wir leben. Die Gestirne sind sehr kleine leuchtende Körper
nahe der Mitte der Kugel. Der Eindruck eines Himmelsgewöl-
bes entsteht nur, weil alle Lichtstrahlen nicht gerade
Linien, sondern Kreise durch den Mittelpunkt der Hohlkugel
sind. Die Theorie verschwand aus den Kiosken, nachdem sie
irrig prophezeit hatte, ein in Russland abgeschossener
Sputnik müsse nach einem Flug von höchstens 13600 Kilometern auf der anderen Seite der Erde, also z.B. in Amerika,
wieder herunterfallen. Den Verfassern war wohl entgangen,
dass ihre Theorie empirisch unwiderlegbar gewesen wäre,
wenn sie nicht nur (wie sie es taten) die Optik, sondern
auch die Kinematik materieller Körper und überhaupt die
ganze physikalische Geometrie der Transformation $r \rightarrow R^2/r$
(R = Erdradius, r = laufender Radius in Polarkoordinaten
um den Erdmittelpunkt) unterworfen hätten. Aber vermutlich
wollten sie das auch nicht, denn ihre Theorie wäre damit
zugleich empirisch unbeweisbar, nämlich von der herrschen-
den Theorie überhaupt empirisch ununterscheidbar geworden.

Als drittes Beispiel kann die Allgemeine Relativitätstheorie dienen. Einstein forderte die allgemeine Kovarianz der Grundgleichungen gegen beliebige topologische
Koordinatentransformationen. Die beiden ersten Beispiele
stützen sich auf solche Transformationen (das zweite auf
eine Transformation mit einer Singularität). Man muss also
fragen, wie Konventionalismus und allgemeine Kovarianz
zusammenhängen.

Ein viertes Beispiel bietet die Hamilton-Jacobische
Fassung der klassischen Punktmechanik an. Da die Newtonsche Bewegungsgleichung, in der die Ortskoordinaten und
die Zeit als Variable vorkommen, eine Differentialgleichung
zweiter Ordnung nach der Zeit ist, kann man neben den 3 n
Ortskoordinaten noch 3 n unabhängige Impulskoordinaten
einführen. Die dynamischen Gesetze sind dann in "kanonischer"

Schreibweise invariant gegen gewisse "kanonische" Transformationen der 6 n Orts- und Impulskoordinaten untereinander.
Insbesondere ist bei beliebiger Hamiltonfunktion die Transformation auf "zyklische" Variable möglich. Sie führt 3 n
konstante Impulse und 3 n linear mit der Zeit wachsende
Ortskoordinaten ein. D.h. sie bildet eine Bewegung mit
beliebigem Wechselwirkungsgesetz auf eine reine Trägheits-
bewegung ab.

        Diese dem Physiker wohlbekannten Fakten müssen erkennt-
nis-theoretisch beim ersten Blick verwirrend wirken. Was
ist nun in unserer Physik Beschreibung realer Erfahrungen
und was ist Konvention?

        Der heutige Physiker würde hierauf wohl antworten:
Die Naturgesetze lassen gewisse Transformationsgruppen zu.
Beschreibungsweisen der Natur, die bei diesen Transformationen ineinander übergehen, sind gleichberechtigt; das ist
der Wahrheitsgehalt des Konventionalismus. Eine Grösse, die
von verschiedenen Standpunkten aus verschieden zu beschreiben ist, ist dann eine wohldefinierte physikalische Grösse,
wenn bekannt ist, wie sie sich bei der Gruppe transformiert.
Naturgesetze sind dann nicht Konventionen, wenn sie sich
formal invariant bei allen Transformationen der Gruppe
("allgemein kovariant") ausdrücken lassen.

        Vielleicht darf ich diese Auffassung, ehe ich sie
abstrakt im Detail erörtere, durch eine anekdotische Erinnerung erläutern. Eines Sommerabends sass ich als Gast
Martin Heideggers mit ihm vor seiner Hütte im Schwarzwald.
Wir betrachteten den Sonnenuntergang, während sich im
Osten soeben der fast volle Mond über die Tannen des Berg-
hangs hob. Heidegger sagte zu mir: "Herr v. Weizsäcker, Sie
dürfen doch eigentlich gar nicht sagen, die Sonne gehe
unter. Sie müssen doch sagen, der Erdhorizont hebe sich."
Ich antwortete, im Sinn der soeben geschilderten Auffassung:
"Ich sage völlig unbefangen, die Sonne gehe unter. Denn ich
weiss, dass ich damit dasselbe meine wie wenn ich sagte,
die Erde drehe meinen Horizont über die Sonne herauf. Wenn
ich auf dem Atlantik, von Amerika kommend, am Heck meines
Schiffs einem Schiff nachsehe, das, von Europa kommend, uns
vor kurzem passiert hat, so sage ich, das andere Schiff
tauche langsam hinter dem Horizont unter, wissend, dass
ein Passagier am Heck jenes Schiffs ebenso über unser
Schiff reden wird, und dass wir beide rechthaben." Hatte
Heidegger recht oder ich?

        Die Schärfe des Problems kommt zum Vorschein, wenn man
fragt, welche Transformationen zulässig sind. Nach Felix

Kleins Erlanger Programm definiert die Auswahl einer Gruppe
jeweils eine Geometrie. Die Mathematiker pflegen einen Raum
zunächst als eine Mannigfaltigkeit von Elementen, die
"Punkte" genannt werden, aufzufassen. Man pflegt ferner
vorauszusetzen, dass alle Punkte einer Geometrie gleichberechtigt sind. D.h. die definierende Gruppe muss jeden
Punkt in einen beliebig gewählten Punkt desselben Raumes
überführen können; nur gewisse Relationen zwischen zwei
oder mehr Punkten sollen invariant bleiben.

Für den Erkenntnistheoretiker der Physik entsteht
schon hier ein Problem. Wenn alle Punkte gleichberechtigt
sind, wie kann man sie dann überhaupt voneinander unterscheiden? Wenn man sie nicht durch jeden von ihnen indivi-
duell anhaftende Eigenschaft unterscheiden kann, wie kann
man s a g e n , welcher Punkt durch eine Transformation
in welchen überführt wird? Der Mathematiker macht sich diese Sache leicht; er "denkt sich", die Punkte irgendwie
unterschieden und bezeichnet. [1])Die Bezeichnung, der "Name"
eines Punkts, ist ein Merkmal, das ihm "von aussen" ange-
heftet wird, das also weder eine "innere", noch eine durch
die Raumstruktur ausgezeichnete Eigenschaft des Punktes,
somit nicht Gegenstand der Geometrie ist. Wie aber soll der
Physiker eine solche gedachte Geometrie real anwenden? Wenn
er einem Gegenstand der Erfahrung, den er als Punkt im
Sinne der Geometrie auffassen will, einen Namen gibt, wie
kann er feststellen, ob dieser Name an dem Punkt haftet,
also ob der Punkt, den er kurz darauf mit eben diesem
Namen bezeichnet, derselbe Punkt ist wie zuvor?

Das Problem ist ein Sonderfall der Frage, wie man
individuelle Gegenstände unterscheiden kann, die unter denselben Begriff fallen. Für Gegenstände des täglichen Lebens
ist eine hinreichende Antwort: sie haben stets auch noch
andere Eigenschaften, die durch ihren gemeinsamen Begriff
nicht determiniert sind, m.a.W. sie fallen stets auch noch
unter andere, und zwar verschiedene Begriffe. Hier sind
zwei Katzen, aber eine weisse und eine schwarze. Hier sind
zwei schwarze Katzen, aber eine grössere und eine kleinere,
usw. Fingiert man jedoch zwei begrifflich ununterscheidbare
Gegenstände, so besagt eine Denktradition (von der Leibniz
und Kant in verschiedener Weise Gebrauch gemacht haben),
sie könnten nicht zur selben Zeit am selben Ort sein, seien
also wenigstens stets durch ihren Ort unterschieden. Wenn
nun aber begrifflich strukturlos gedachte Orte selbst, eben
die Punkte, die Gegenstände sind, die man unterscheiden
will, so wäre es zirkelhaft, sie durch den Ort zu unter-

scheiden, an dem sie sich befinden.

Die übliche Antwort ist wohl: Begrifflich gleichartige
Gegenstände, im Idealfall also Punkte, kann man nur demon-
strativ, durch Hinzeigen unterscheiden. Dadurch wird nunmehr das zeigende Subjekt als unerlässliche Voraussetzung
des Sinns der verwendeten Begriffe in die Erkenntnistheorie
der Physik eingeführt. Ich halte diesen Schritt in der Tat
für fundamental und für unvermeidlich. Dies ist hier zunächst eine blosse Behauptung. Die folgenden Abschnitte
sollen die semantisch konsistente Einführung der physikalischen Geometrie schrittweise diskutieren, geleitet von
der progressiven Erkenntnis der konventionalistischen
Freiheit.

6. DINGLERS OPERATIVE BEGRÜNDUNG DER EUKLIDISCHEN GEOMETRIE

Hugo Dingler (1938)[2]) hat in Kenntnis der Kraft der konventionalistischen Argumente Poincarés versucht, die hierarchische Überordnung genau der euklidischen Geometrie über
die empirische Physik zu retten. Er behauptet nicht mehr
eine ontologische Wahrheit dieser Geometrie für einen
physikalischen Gegenstand "Raum", sondern ihre operative
Notwendigkeit für einen methodisch eindeutigen Aufbau der
Physik. In diesem Sinne meint er Kants Gedanken zu verwirk-
lichen, Geometrie gehöre zu den Bedingungen der Möglichkeit
der empirischen Physik. Lorenzen hat diese fast verscholle-
nen Gedanken Dinglers wieder aufgenommen und zum Programm
einer "Protophysik" ausgebaut. Diese Thesen Lorenzens könnten adäquat nur im Zusammenhang mit seinem Entwurf einer
Protologik, also allgemein einer Wiederherstellung des
Hierarchismus auf operativer oder diskursiver Grundlage
erörtert werden. Das kann im gegenwärtigen Aufsatz nicht
geschehen. Daher sei hier nur ein m.E. entscheidender
Grundgedanke Dinglers besprochen.

Eine Geometrie im Sinne des Erlanger Programms wird
definiert durch eine Gruppe von Abbildungen der zugrunde-
gelegten Punktmannigfaltigkeit auf sich. Dabei fordert man
zwar, dass jeder Punkt in jeden anderen überführt werden
kann (Homogenität des Raumes). Aber eine Teilmannigfaltigkeit von Punkten (eine "Figur") soll nicht in jede gleich-
mächtige Teilmannigfaltigkeit überführt werden können.
Figuren, die durch die Gruppe ineinander überführt werden
können, heissen gleich im Sinne der betr. Geometrie. Die
engsten Gruppen und damit die Einteilung in die kleinsten
Klassen, die in der Geometrie betrachtet zu werden pflegen,

halten eine Metrik invariant, speziell die euklidische
Gruppe.

Man kann sagen, dass die euklidische Gruppe den
klassischen Schnitt zwischen Geometrie und Physik (s. Abschnitt 1) überhaupt erst definiert hat. Sie überführt
"kongruente" Körper ineinander ohne Rücksicht auf ihren Ort
und ihre Orientierung. Physikalisch hingegen lehrt die ele-
mentare Erfahrung den Wesensunterschied von Richtungen wie
z.B. oben und unten. Die griechische Wissenschaft kannte
die Denkmöglichkeit, die Auszeichnung von Orten und Richtungen auf die Relation zu bestimmten Körpern (z.B. zur
Erde) oder zum Weltall zurückzuführen. Einerlei welches
Modell der Physik und Kosmologie man wählte, dies modifi-
zierte jedenfalls nicht die Geometrie, insofern diese
kongruente, aber verschieden situierte und vielleicht qua-
litativ verschieden beschaffene Körper als äquivalent behandelt. Aus dieser Erfahrung der Existenz von Eigenschaften aller Körper, die von allen ihren übrigen empirischen
Unterschieden unabhängig waren, erwuchs die Vorstellung
einer hierarchisch übergeordneten Wissenschaft von Körpern
überhaupt, die nunmehr auch nicht auf Erfahrung gegründet
schien, eben der Geometrie.

Der Empirismus des 19. Jahrhunderts musste demgegenüber die Angabe derjenigen Erfahrung verlangen, die diese
Überordnung rechtfertigte. Helmholtz fand sie in der
Existenz frei verschieblicher starrer Körper. Diese Forderung reichte aber nur aus, um eine metrische Geometrie
mit konstantem Krümmungsmass zu begründen. Dingler verwendete eine gruppentheoretisch weitergehende Forderung, die
am bequemsten in der für ihn ohnehin zentralen operativen
Fassung dargestellt wird. Wenn Glas- oder Metallschleifer
eine präzise (euklidische) Ebene herstellen wollen, so
schleifen sie drei Körper wechselweise aneinander ab. Zwei
Körper,aneinander abgeschliffen, würden kongruente Grenz-
flächen konstanter Krümmung erzeugen. Wenn beide kongruent
auf eine dritte Grenzfläche passen sollen, so muss die
Krümmung Null sein.

Dingler argumentierte nun im wesentlichen so: Wenn wir
geometrische Messungen ausführen wollen, müssen wir zunächst die geometrischen Eigenschaften der Messgeräte festlegen. Dies muss in eindeutiger und darum unstreitig wieder-
erkennbarer Weise geschehen und wird durch eine Reihe von
real ausführbaren Operationen erreicht, von denen die soeben geschilderte Herstellung einer Ebene wohl die wichtigste
ist. Diese Operationen garantieren, dass die Messinstrumente

mit derjenigen Genauigkeit, mit der die eindeutigen Vor-
schriften bei ihrer Herstellung befolgt wurden – prinzipiell
also mit beliebiger Genauigkeit – einer ganz bestimmten
Geometrie genügen, und zwar der euklidischen. Folglich
müssen auch alle Objekte, die man mit diesen Geräten aus-
misst, kraft dieser Messungen geometrisch durch Eigenschaften charakterisiert werden, die notwendigerweise nur innerhalb dieser Geometrie scharf definiert sind und darum den
Gesetzen eben dieser Geometrie genügen. Also ist durch eine
reine Operationsvorschrift a priori gewiss, dass jeder in
der Erfahrung mögliche geometrische Sachverhalt der euklidischen Geometrie genügen wird. Wenn jemand nun z.B. in
Lichtstrahlendreiecken Winkelsummen ungleich zwei Rechten
findet, so ist a priori gewiss, dass dies nur die Deutung
zulässt, dass Lichtstrahlen keine Geraden sind. Man sieht
sofort, dass keines unserer vier konventionalistischen
Argumente, so wie es bisher vorgebracht ist, diesem Einwand standhält. In keinem von ihnen ist erwogen, wie man
es macht, geometrische Messinstrumente herzustellen (zum
dritten Beispiel, der ARTh, vgl. Abschnitt 9). Das am
Ende des Abschnitts 5 erörterte Problem der Bezeichnung
von individuellen Punkten ergänzt Dingler somit durch eine.
Erwägung der operativen Kennzeichnung geometrischer Begriffe, also von Figurenklassen. Nur in Bohrs Diskussion des
Messprozesses (Abschnitt 4 Ende) findet sich eine Analogie,
und Bohr ist ja in der Tat zu dem Resultat gekommen, dass
jede Messung klassisch (also vermutlich, obwohl Bohr m.W.
darüber nichts behauptet hat, auch euklidisch) beschrieben
werden muss.

   Als ich Dinglers Gedanken um 1935 kennenlernte, sah
ich sofort ihre Kraft und reagierte auf sie doch so, wie
fast alle Physiker zuvor und danach auf sie reagiert haben:
Einsteins Einführung der Riemannschen Geometrie in die
Physik kann doch durch solche methodologischen Argumente
nicht als falsch erwiesen werden. Also musste Dinglers
Argument auch methodologisch einen Fehler enthalten. Als
ich diesen Fehler (wie ich auch heute meine, zutreffend)
lokalisiert hatte, schrieb ich darüber einen Aufsatz und
besuchte Dingler zu einer achtstündigen, natürlich ergeb-
nislosen Diskussion, deren Hauptargumente ich hier nach
meiner (vermutlich subjektiv gefärbten) Erinnerung schil-
dere.

   Die Analogie mit Bohr gab mir einen Wink. Bohr argumentierte, wenn auch sehr viel weniger präzisiert als
Dingler, ein Messinstrument müsse raumzeitlich beschreib-

bar sein (sonst kann man es nicht wahrnehmen) und streng
kausal funktionieren (sonst kann man aus der Ablesung nicht
auf das Messobjekt schliessen); nur in der klassischen
Physik seien de facto aber Raum-Zeit-Beschreibung und
Kausalforderung vereinbar, und darum müsse man ein Messinstrument klassisch beschreiben. Die oft erörterten Probleme,
die dieses Argument aufwirft, bespreche ich hier nicht. Ich
unterscheide, nur zum Vergleich mit Dingler, zwei denkbare
Interpretationen der Intention Bohrs, eine falsche und eine
richtige. Bohr argumentiert n i c h t so: "Jedes Messinstrument genügt der klassischen Physik. Also definiert es
alle physikalischen Eigenschaften der Messobjekte im Einklang mit der klassischen Physik. Also gilt die klassische
Physik für alle messbaren Objekte." Er argumentiert vielmehr so: "Jedes Messinstrument ist nur soweit zur Messung
tauglich, als man es klassisch beschreiben kann. Nun gilt
aber für die Messobjekte, jedenfalls im atomaren Bereich,
nicht die klassische Physik, sondern die Quantentheorie.
Deshalb lassen sich diese Objekte nicht wie makroskopische
Körper genähert objektivieren; das ist die Komplementari-
tät. Die Messgeräte also müssen makroskopische Körper sein,
und wo die Näherung, in der wir sie mit den klassischen
Begriffen beschreiben, zusammenbricht, sind sie eben nicht
mehr als (ideale) Messinstrumente tauglich."

Analog argumentierte ich nun gegen Dingler. Dingler
behauptet, sein Argument gelte a priori im Sinne Kants,
d.h. unabhängig von jeder speziellen Erfahrung, abhängig
nur davon, dass überhaupt Erfahrung (räumlicher Art) möglich ist. Dann müsste es aber auch gelten, wenn man die
Hypothese machte, die speziellen Erfahrungen seien so, dass
alle Körper und das Licht sich verhalten wie Körper und
Strahlen einer bestimmten nichteuklidischen Geometrie, die
nur im Kleinen (praktisch: im täglichen Erfahrungsbereich
der Menschen) in hinreichender Näherung euklidisch approximiert werden kann. Unter dieser Voraussetzung müsste aber
Dinglers operative Herstellung euklidischer Ebenen für
hinreichend grosse Werkstücke (oder hinreichend genaue Realisierung) physisch in vorhersagbarer und reproduzierbarer
Weise mißraten. Ich nahm z.B. an, sie müssten sich bei den
zum Anpassen und Schleifen nötigen Bewegungen so deformie-
ren, dass sie total abgeschliffen würden, ohne je zur
Deckung zu kommen. Vielleicht läge noch eher eine Annahme
nahe, nach der das Schleifen zwar gelingt, aber ein auf einer so hergestellten Ebene euklidisch gezeichnetes Dreieck
bei der Bewegung seine (durch Messungen im Kleinen an drei

verschiedenen Orten messbare) Winkelsumme ändert. Als die
Diskussion so weit fortgeschritten war, antwortete Dingler
meiner Erinnerung nach nur, das von mir unterstellte Verhalten der Körper sei "eine phantastische Behauptung", für
die ich die Beweislast trage. Damit glaubte (und glaube)
ich die Debatte definitiv gewonnen zu haben. Denn mein
Argument bedarf nicht der empirischen Richtigkeit, sondern
nur der Denkbarkeit dieser Annahme über das Verhalten
physischer Körper. Damit ist eine spezielle Erfahrung
denkbar, die Dinglers Forderung in einer exakt angebbaren
Grössenordnung unerfüllbar macht. Also kann man nicht
a priori einsehen, dass Dinglers Annahme von jeder speziellen Erfahrung unabhängig sein muss, entgegen seinem Anspruch. Etwas unfreundlicher formuliert: Eine physikalische
Hypothese ist nicht deshalb unmöglich, weil sie vom Standpunkt einer zum Zweck ihrer Vermeidung eingeführten Metho-
dologie aus phantastisch erscheint.

7. SEMANTISCHE KONSISTENZ

Wir schalten einen Abschnitt wissenschaftstheoretischer
Besinnung ein und vergegenwärtigen uns zunächst den bis
hierher erreichten Stand in der Frage nach den Grundlagen
der physikalischen Geometrie. Wir befinden uns im Versuch
einer gruppentheoretischen Präzisierung der durch den
Konventionalismus erzeugten Fragestellung. Wir haben zunächst die Annahme akzeptiert, es gebe eine Mannigfaltigkeit von Punkten, genannt "der Raum", die untereinander
naturgesetzlich äquivalent, aber durch Aufweisung unterscheidbar sind. Wir wissen schon, dass wir diese Mannigfaltigkeit, wenn wir die Relativitätstheorie übernehmen,
durch die raumzeitliche Mannigfaltigkeit von punktuellen
Ereignissen werden ersetzen müssen, die Minkowski "die
Welt" nannte (Abschnitt 8). Ferner sollten wir darauf
vorbereitet sein, die Annahme einer solchen Punktmannigfaltigkeit quantentheoretisch zu kritisieren und vielleicht
zu begründen (Abschnitt 10). Unsere gegenwärtige Frage
ist, ob und, wenn ja, wodurch eine Gruppe von Abbildungen
des Raumes auf sich ausgezeichnet ist, welche eine physikalische Geometrie, also eine Äquivalenzrelation von
Figuren, definieren würde.
      Historisch haben wir eine solche Auszeichnung in
Gestalt der euklidischen Geometrie schon vorgefunden. Wir
haben die kritische Rückfrage nach der Rechtfertigung einer
solchen Auszeichnung (sei es der euklidischen, sei es einer

anderen Geometrie) gestellt und drei sukzessive Antworten,
die wir erhielten, als unzureichend befunden: die Rechtfertigung durch Evidenz, durch direkte Erfahrung und durch
hierarchistische Festlegung von Bedingungen der Möglichkeit
von Erfahrung. Damit haben wir aber auch am Beispiel der
Geometrie ein Stück Geschichte der Wissenschaftstheorie
nachvollzogen. Wie kann man überhaupt der Wahrheit eines
allgemeingültigen Satzes gewiss sein? Die Geschichte der
Reflexion auf diese Frage bietet uns, roh gesprochen,
gerade die vier Antworttypen: naive Selbstverständlichkeit,
Berufung auf Evidenz, Bestätigung durch Erfahrung, Reflexion auf die Bedingung der Möglichkeit von Erfahrung. Man wird
sagen dürfen, dass alle vier Verifikationsweisen, jedenfalls so, wie sie bisher präsentiert worden sind, aus
allgemein einsichtigen Gründen scheitern müssen. Naive
Selbstverständlichkeit einer Meinung ist überhaupt kein
Argument für diese, sondern sie ist der Zustand, ehe das
Bedürfnis nach Argumenten gefühlt wird. Ihre Selbstvertei-
digung gegen kritische Rückfragen ist die Berufung auf
Evidenz. Berufung auf Evidenz ist aber dort kein Argument,
wo jemand da ist, der diese Evidenz nicht in sich erlebt.
Das mag daran liegen, dass er etwas nicht sieht, was ihm,
wenn er es sähe, evident wäre; es kann aber auch daran
liegen, dass er ein Gegenargument oder Gegenbeispiel sieht.
Der im Wort Evidenz steckende Begriff des Sehens erscheint
nun der empiristischen oder sensualistischen Reflexion als
eine Metapher. Sie sucht Evidenz auf sinnliche Evidenz ein-
zuschränken und die Realwissenschaften wie die Physik auf
diese zu begründen. Der von Popper wieder hervorgehobene
Einwand hiergegen ist, dass die Gesetze der Physik allgemeingültig sein sollen, und dass sie als allgemein nicht
durch eine ihrem Wesen nach unvollständige Aufzählung von
Fakten begründet werden können. Kants Begründung allgemeiner Gesetze als notwendige Bedingungen der Erfahrung
bleibt von diesem Einwand nur dann unbetroffen, wenn die
Notwendigkeit dieser Bedingungen selbst evident ist. Dass
ihr diese Evidenz jedenfalls in der von Kant oder Dingler
gewählten hierarchistischen Fassung fehlt, haben wir am
Ende der Abschnitte 3 und 6 wohl gesehen. Wir lassen dabei,
wie überhaupt in diesem Aufsatz, das Problem logischer oder
struktureller Evidenz beiseite.

Wir könnten soweit der Popperschen These folgen, dass
allgemeine, für die Erfahrung gültige Gesetze nicht veri-
fiziert werden können. Nicht so klar ist, ob sie wenigstens
falsifiziert werden können. Unsere Gegenbeispiele gegen

Kant und Dingler standen auf einer höheren Abstraktions-
stufe als derjenigen der empirischen Falsifikation. Es
wurde nicht behauptet, eine Erfahrung liege vor, die mit
dem euklidischen Charakter der reinen Anschauung oder der
Herstellbarkeit euklidischer Ebenen unvereinbar sei, sondern nur, Hypothesen über die Anschauung oder über das Verhalten von Körpern seien möglich, welche, wenn sie wahr
wären, die Realisierung der Erwartungen Kants und Poppers
faktisch ausschliessen würden. Im übrigen widersteht weder
der Verifikations- noch der Falsifikationsglaube der Kritik,
die der Konventionalismus an den Prämissen beider übt.

Die durch Th. S. Kuhn (1962) eingeleitete neuere Entwicklung der Wissenschaftstheorie lässt diese bisher ungelösten Probleme mehr oder weniger auf sich beruhen und
studiert zunächst die geschichtliche Entwicklung der
Wissenschaft. Diese Wendung kann man als eine legitime
Kritik der Wissenschaftstheorie an ihrem eigenen Hierarchismus verstehen. Auch die Wissenschaftstheorie ist der
Wissenschaft, deren Theorie sie ist, nicht hierarchisch
vorgeordnet; sie ist durch den historischen Gang der
Wissenschaft korrigierbar. Anders gesagt: Wenn Erfahrung
die Grundlage unseres Wissens genannt wird, so wird man
wohl erst durch Erfahrung wissen können, was Erfahrung
ist; und der hierfür relevante Erfahrungssbereich ist
zunächst einmal die Geschichte der Wissenschaft. Kuhn
beschreibt diese Geschichte als eine Abfolge von Paradigmen, deren jedes eine Phase normaler Wissenschaft be-
herrscht und in einer wissenschaftlichen Revolution durch
ein neues abgelöst wird. Heisenberg (1948,1972) hatte
schon vor Kuhn denselben historischen Hergang als eine
Abfolge abgeschlossener Theorien beschrieben, deren jede
die früheren umfasst und sie auf einen Bereich genäherter
Geltung einschränkt.

Uns hat der Unterschied zwischen Kuhn und Heisenberg
zu interessieren, also der Unterschied zwischen einem
zur Problemlösung brauchbaren paradigmatischen Verfahren
und einer einen Erfahrungsbereich in hinreichend guter
Näherung beschreibenden Theorie. Für Kuhns These, dass
paradigmatische Verfahren allgemeiner verwendet werden
als die zu ihrer Interpretation benützten theoretischen
Regelsysteme, gibt es zahlreiche historische Beispiele.
Aber sind unsere hauptsächlichen Beispiele physikalischer
Geometrie, nämlich die in der klassischen Physik benutzte
euklidische, die in der SRth eingeführte Minkowskische
und die in der ARth verwendete Riemannsche, nicht abge-

schlossene Theorien im Sinne Heisenbergs? Sie sind wohldefinierte mathematische Theorien. Aber ihre Anwendung
auf die empirische, technische Wirklichkeit setzt voraus,
dass man weiss, mit welchen Phänomenen dieser Wirklichkeit man die in ihnen verwendeten mathematischen Begriffe,
beginnend mit Begriffen wie Punkt, Gerade, Abstand, iden-
tifizieren soll. D.h. sie setzt gerade voraus, dass man
nicht im Problem des Konventionalismus steckt.

Ich schlage nun vor, auch dieses Problem bis auf
weiteres geschichtlich zu betrachten. Auch die Deutungen
der mathematischen Theorien geschehen durch Paradigmen,
und eben diese Paradigmen sind im allgemeinen die Rechtfertigung der Einführung der Theorien. Eine Theorie eines
Phänomenbereichs geht stets von einem Vorverständnis der
Phänomene aus. Wenn die Theorie volle mathematische
Gestalt angenommen hat, so kann man in ihr den mathematischen Formalismus unterscheiden von der physikalischen
Deutung der im Formalismus benutzten Begriffe. Die Deutung
bedient sich des Vorverständnisses, z.B.: Seien $x_1 x_1 z_1$ und
$x_2 y_2 z_2$ die cartesischen Koordinaten zweier Punkte an
Körpern, so sei $r = \sqrt{(x_1-x_2)^2 + (y_1-y_2)^2 + (z_1-z_2)^2}$ ihr
Abstand. Was Abstand ist, wissen wir alle schon.

Hier tritt nun ein zentrales wissenschaftstheoreti-
sches Problem auf, das ich als das Problem der semantischen Konsistenz bezeichnen möchte. Das Vorverständnis
liefert die Semantik, die physikalische Bedeutungserfül-
lung der mathematischen Theorie. Die so gedeutete Theorie
aber beschreibt, wenigstens in gewissen Aspekten, eben
dieselben Phänomene, in denen sich das Vorverständnis
bewegt. Ist diese theoretische Beschreibung der Phänomene
mit dem Vorverständnis vereinbar? D.h. ist die Theorie,
deren mathematische Konsistenz vorausgesetzt sei, in dem
Sinne auch semantisch konsistent, dass der Gesamtkom-
plex der gedeuteten Theorie, also die mathematische Theorie
zusammen mit dem Vorverständnis, widerspruchsfrei ist?
Diese Frage lässt sich oft nicht unmittelbar entscheiden,
jedenfalls dann nicht, wenn das Vorverständnis selbst
nicht theoretisch formuliert war. Es kann vorkommen, dass
die Theorie nunmehr das Vorverständnis präzisiert. D.h.
man gewöhnt sich an, das Vorverständnis im Einklang mit
der Theorie auszusprechen und alle dazu nicht passenden
Sprech- und Vorstellungsweisen zu verbannen. In diesem
Sinne hat die euklidische Geometrie das Vorverständnis
der räumlichen Anschauung präzisiert (vgl. Ende von Abschnitt 4). Es kann aber auch vorkommen, dass die Theorie

das Vorverständnis explizit korrigiert. Das klassische
Beispiel dafür ist Einsteins Kritik des Begriffs der
Gleichzeitigkeit entfernter Ereignisse (Abschnitt 8). Jedenfalls aber wird man die semantische Konsistenz für eine
abgeschlossene Theorie fordern.

Diese Forderung erweist sich jedoch als zweideutig.
Sie lässt eine engere und eine weitere Fassung zu. Dies
hängt damit zusammen, dass Theorien Phänomene nicht nur
historisch beschreiben, sondern erklären. Die sehr kom-
plexe wissenschaftstheoretische Debatte über den Erklä-
rungsbegriff soll hier nicht aufgerollt werden. Der Unterschied von Beschreiben und Erklären werde vielmehr nur am
Beispiel der euklidischen Geometrie erläutert. Wenn die
babylonische Geometrie den pythagoräischen Lehrsatz wie
ein faktisch geltendes Naturgesetz kannte, so wollen wir
sagen, dass sie die Natur mit Hilfe dieses Gesetzes be-
schrieb. Wenn ein griechischer Mathematiker, etwa Pythagoras, diesen Lehrsatz aus wenigen Axiomen herleitete und
in diesem Sinne bewies, so sagen wir, er habe die von den
Babyloniern beobachteten Phänomene erklärt. Beschreiben
und Erklären sind, so gefasst, Relativbegriffe. Man kann
auch sagen, dass die Babylonier die Phänomene, die sie
im einzelnen beschrieben, durch den empirisch fundierten,
aber als allgemeingültig postulierten Lehrsatz erklärten.
Andererseits kann man sagen, dass die Griechen mit den
Prinzipien, aus denen heraus sie erklärten, also z.B. mit
den Axiomen der euklidischen Geometrie, wieder nur gewisse
sehr allgemein verbreitete Phänomene beschrieben. In dieser
Verschieblichkeit der Terminologie verbergen sich die Probleme des Sinns der Begriffe "Begriff" und "Gesetz". Jedenfalls aber statuiert eine Erklärung, so wie wir hier das
Wort gebrauchen, stets die Notwendigkeit eines Zusammenhangs, der ohne sie nur beschrieben werden könnte; und
Notwendigkeit ist selbst nur relativ auf andere, selbst
letzten Endes entweder als evident erlebte oder nur beschriebene Zusammenhänge. Fassen wir nun das Vorverständnis einer Theorie als die Beschreibung eines Phänomenbereichs auf, so kann die Theorie einen Teil dieser Beschreibung erklären (wie der pythagoräische Beweis die allgemeine
Gültigkeit des pythagoräischen Lehrsatzes erklärt), einen
Teil wird sie selbst lediglich beschreiben, eventuell mit
präziseren Begriffen (z.B. indem sie die Existenz der
Fundamentalgebilde, wie Punkte, Geraden, Körper postuliert),
einen dritten Teil wird sie als ausserhalb ihres Gegen-
standsbereichs liegend auf sich beruhen lassen (z.B. den

physischen Unterschied zwischen den Raumrichtungen).

Als semantisch konsistent im engen Sinne kann man nun eine Theorie bezeichnen, deren Vorverständnis, soweit es von der Theorie selbst erklärt wird, mit dieser Erklärung verträglich ist. Z.B. dürfen im Vorverständnis der euklidischen Geometrie keine Dreiecke vorkommen, für welche empirisch die Verletzung des Satzes des Pythagoras oder des Winkelsummensatzes behauptet wird. Wir wollen sagen, eine Theorie sei semantisch konsistent im engeren Sinne, wenn in ihr keine semantischen Inkonsistenzen bekannt geworden sind.

Semantisch konsistent im weiteren Sinne werden wir eine Theorie nennen wollen, in der gar keine semantischen Inkonsistenzen auftreten können. Diese Forderung lässt wieder zwei Interpretationen zu. Die erste Interpretation entspricht Heisenbergs Begriff der abgeschlossenen Theorie. Die Theorie sei als mathematisch widerspruchsfrei vorausgesetzt, und es sei verfügt, dass das Vorverständnis nur im Einklang mit der Theorie interpretiert werden darf. Dann ist die semantische Konsistenz durch Verfügung gesichert. Damit ist aber vom eigentlichen Problem der semantischen Konsistenz abgesehen. Man kann ja nicht durch Verfügung sichern, dass die Phänomene eine Beschreibung gemäss dieser Theorie überhaupt unbegrenzt zulassen. Dies verfügen zu wollen, war Dinglers Fehler. Heisenberg spricht daher konsequent vom Geltungsbereich einer abgeschlossenen Theorie. Dieser umfasst gerade die Phänomene, die sich der Verfügung fügen. An welcher Stelle der Geltungsbereich aufhört, das kann nur weitere Erfahrung lehren oder die Erklärung dieser weiteren Erfahrung in einer umfassenden abgeschlossenen Theorie. Dabei ist die nun nicht mehr mit der Theorie vereinbare Erfahrung eine Art Erweiterung des Vorverständnisses über die bisherige Theorie hinaus. Die neue Erfahrung wird im allgemeinen mit Hilfe der Begriffe der bisherigen Theorie formuliert und führt gerade dann auf Widersprüche. Erst die neue Theorie verändert mit den neuen Gesetzen, in denen die Begriffe vorkommen, den Sinn der Begriffe so, dass die Widersprüche verschwinden ("sich als scheinbar erweisen"). In diesem Sinne ist die alte Theorie Vorverständnis der neuen. Im allgemeinen wird die neue Theorie einen Teil dessen erklären, was die alte nur beschreibt. Heisenbergs abgeschlossene Theorien erweisen sich damit als semantisch konsistent im engeren Sinne, solange sie als wahr gelten, und, wenn sie überholt sind, als semantisch konsistent im weiteren Sinne nur

bezüglich ihres Geltungsbereichs, ausserhalb davon als
semantisch inkonsistent.

Die zweite und eigentliche Interpretation der semantischen Konsistenz im weiteren Sinne müsste eine Theorie
bezeichnen, in der solche Widersprüche nicht mehr auftreten können. Dies könnte höchstens in einer letzten,
endgültigen Theorie der Fall sein. Gegenüber einer historischen Abfolge von Theorien bleibt es eine regulative
Idee, deren Diskussion den Rahmen dieses Aufsatzes über-
schreitet.

Wir können in der jetzt eingeführten Sprechweise unsere bisherigen Probleme beschreiben. Der naive Empirismus
übersieht, dass jede Erfahrung ein Vorverständnis benutzt,
das selbst schon Begriffe, also Theorieelemente enthält
und entsprechend kritisierbar ist. Der Hierarchismus er-
kennt dieses Vorverständnis als solches und versucht, es
ein für allemal festzulegen. Er übersieht die Möglichkeit
seiner nachträglichen Korrektur. Diejenigen Züge des Vorverständnisses, deren Abänderung man für unmöglich hält,
sofern überhaupt Erfahrung stattfinden soll, nennt man
Bedingungen der Möglichkeit von Erfahrung. Unsere Auffassung besagt, dass man von solchen Bedingungen sinnvoll
sprechen kann, dass aber ihre Formulierung, ja ihr gesam-
ter Inhalt durch nachträgliche Prüfung der semantischen
Konsistenz verändert werden kann. Der Konventionalismus
schliesslich hebt das Problem aus seinem geschichtlichen
Zusammenhang heraus und wäre daher vielleicht erst in
einer im weiteren Sinne semantisch konsistenten endgültigen Theorie vollständig diskutierbar. Wir beschränken
uns hier auf die Beispiele aus Abschnitt 5.

Das erste Beispiel (Gauss' Dreiecksmessung) formuliert nur eine Aufgabe. Es unterstellt, innerhalb des
kombinierten Vorverständnisses der euklidischen Geometrie
und der Theorie geradliniger Lichtausbreitung werde ein
empirischer Widerspruch konstatiert. Nach unserer jetzigen
Auffassung muss erst eine Theorie gefunden werden, die
diese Phänomene widerspruchsfrei beschreibt, und diese
Theorie wird dann das Vorverständnis korrigieren. Ohne
solche Theorie ist das Beispiel nicht diskutierbar.

Das zweite Beispiel (Hohlwelt) ist umgekehrt innerhalb einer Theorie (z.B. der klassischen Physik) formuliert. Hält man an dem Vorverständnis fest, dass ein Abstand die Anzahl von Malen misst, in der ein Einheitsmass-
stab angelegt werden muss, um ihn auszumessen, so ist
gemäss dieser Definition die Beschreibung der Welt als

Hohlwelt semantisch inkonsistent, denn sie fordert, dass
sich auch die Masstäbe gemäss der Transformation $r \rightarrow R^2/r$
ändern. Im übrigen zeigt das Beispiel nur, dass in einer
im engeren Sinne semantisch konsistenten Theorie Umbenennungen möglich sind, die gerade deshalb empirisch unwiderlegbar sind, weil die neuen Namen durch explizite Definition aus den alten hervorgehen. Der Widerstand, den wir
gegen solche Umbenennungen empfinden, hängt damit zusammen, dass Sprache und Vorverständnis innerhalb unserer
Geschichte nie in ihrem Beitrag zu einer einzigen, wenn
auch noch so umfassenden Theorie aufgegangen sind. Dies
wird deutlich an Heideggers Äusserung über den Sonnenuntergang. Meine relativistische Antwort war ungenau. Die
beiden Schiffe zwar sind auch mechanisch äquivalent, da
ihre Bewegungen durch eine Raumspiegelung auseinander
hervorgehen. Für die klassische Dynamik ist jedoch die
kopernikanische Beschreibung des Sonnenuntergangs, die
Heidegger mir imputierte, in der Tat ausgezeichnet, da
in rotierenden Bezugssystemen "Scheinkräfte" auftreten;
das Wort "Schein" bezeichnet hier das Empfinden semantischer Inkonsistenz. Freilich kann die sprachliche Freiheit in der Formulierung des Vorverständnisses so erwei-
tert werden, dass auch die Folgen solcher Übergänge zu
beschleunigten Bezugssystemen korrekt mitgedacht werden.
Auf der anderen Seite umfasst die übliche Sprechweise,
die Heidegger für sich in Anspruch nahm, die volle Beziehung des wahrnehmenden Subjekts zum sogenannten physischen Vorgang, also das eigentliche Phänomen. Auf diese
Trennung des physikalisch objektivierten Vorgangs vom
Phänomen wies Heidegger hin. Aber gerade auf diese seine
Intention zielte meine Antwort. Ich war der Meinung, dass
das Subjekt sich anderen Subjekten gegenüber relativieren
kann und soll. Deshalb sprach ich von einem Passagier auf
dem anderen Schiff, der vielleicht - was ich nicht sagte -
mein Freund ist. Mit veränderter Bewusstseinslage verändern
sich auch die Phänomene. Doch gehört der Bewusstseins-
wandel durch die Wissenschaft (die Zusammengehörigkeit
von "Du" und "Gestalt") zu den vielen in diesem Aufsatz
nicht mehr erörterten Problemen.

Das dritte Beispiel besprechen wir im 9. Abschnitt.
    Das vierte Beispiel (Hamilton - Jacobi - Theorie)
ist ähnlich gebaut wie das zweite, aber physikalisch sinn-
voller. Der Übergang zu den zyklischen Koordinaten ist
gleichbedeutend der Lösung der Bewegungsgleichungen. Es

charakterisiert jede Lösung durch ihre Integrationskon-
stanten. Die klassische Punktmechanik bedarf also eines
Vorverständnisses, um freie von Wechselwirkungsvorgängen
zu unterscheiden, und ihre Invarianz gegen kanonische
Transformationen zeigt, dass sie, für sich genommen, dieses
Vorverständnis selbst nicht erklären kann. Wie in den Abschnitten 9 und 10 näher besprochen, hat dieses Vorverständnis mit der Zerlegbarkeit in Einzelobjekte (hier
Massenpunkte) zu tun.

    Zusammenfassend kann man vielleicht sagen, dass der
Konventionalismus zwei ziemlich verschiedene Problemklas-
sen andeutet. Im Bereich des Vorverständisses weist er auf
den konventionellen oder historisch relativen Charakter
der Sprache, auf die Möglichkeit verschiedener Sprachspiele
hin. Dies gehört in die Sprachphilosophie und kann vielleicht überhaupt nicht mit mathematischer Strenge erörtert
werden. Wenn man lange genug über solche Fragen redet und
sensibel und guten Willens ist, versteht man einander am
Ende ein Stück weit und kann eine Sprache entwickeln, in
der dieses Verständnis wirksam wird. Im Bereich der mathematischen Theorie reduziert sich der Konventionalismus auf
die Invarianz der Gesetze gegen Transformationsgruppen.
Welche Gruppen das sind, müssen wir nunmehr mit Hilfe des
Vorverständnisses der heutigen Physik zu formulieren suchen.

8. SPEZIELLE RELATIVITÄTSTHEORIE

Das Vorverständnis der SRth enthält unter anderem die euklidische Geometrie und die Newtonsche Mechanik. Beide
können hier als abgeschlossene physikalische Theorien
betrachtet werden, nach deren semantischer Konsistenz
gefragt werden darf.

    Eine physikalische Axiomatik der euklidischen Geome-
trie wird zweckmässigerweise die Helmholtz-Dinglerschen
Operationen an starren Körpern zugrundelegen. Sie begründet so die 6-parametrige euklidische Gruppe der reell-
dreidimensionalen Rotationen und Translationen. Die Theorie
ist dann semantisch konsistent im engeren Sinne. Ihr Vorverständnis setzt jedoch die Existenz starrer Körper voraus.
Dies ist erstens eine Idealisierung. Die Phänomene zeigen
die Existenz starrer Körper nur genähert. Sie rechtfertigen
nicht eo ipso die Fiktion beliebig genauer Annäherung an
das Ideal. Man muss also auf die Möglichkeit vorbereitet
sein, einen begrenzten Geltungsbereich der Theorie zu
entdecken. Zweitens wird die Existenz starrer Körper von

der Geometrie nur im Vorverständnis benutzt, aber nicht theoretisch erklärt. Die statistische Physik des späten 19. Jahrhunderts hat schrittweise entdeckt, dass die Anwendung der klassischen Mechanik auf das Innere der Körper auf Schwierigkeiten führt, die vermutlich prinzipiell unüberwindlich sind. Jedenfalls hat erst die Quantentheorie dieses Problem gelöst (vgl. Abschnitt 10).

Die klassische Mechanik fügt, wie man unter der gruppentheoretischen Fragestellung gegen Ende des 19. Jahrhunderts erkannte (L. Lange), eine vierparametrige Erweiterung zur euklidischen Gruppe hinzu, aus Transformationen bestehend, die die Zeit enthalten. Die einparametrige Untergruppe der Zeittranslationen, die die Homogenität der Zeit ausdrückt, hat man meist als Formulierung der Annahme, dass dieselben Naturgesetze immer gelten, leicht akzeptiert. Hingegen enthalten die "eigentlichen Galilei-Transformationen", die Inertialsysteme ineinander transformieren, das spezielle Relativitätsprinzip, das viele philosophische Diskussionen wachgerufen hat. Historisch sind diese Diskussionen in zwei Phasen abgelaufen, die man als die Phase vor Einstein und die Phase nach Einstein unterscheiden kann. Vor Einstein erschien das Relativitätsprinzip nur dann als ein allgemeines Naturprinzip begründet, wenn man die klassische Mechanik als die fundamentale Wissenschaft von der Natur ansah; man kann diese Prämisse auch als das mechanische Weltbild bezeichnen. Etwa unter dieser Prämisse wurde die Relativität der Bewegung z.B. von Leibniz (gegen Clarke, d.h. gegen Newton), von Kant (in den Metaphysischen Anfangsgründen der Naturwissenschaft) und von Mach (ebenfalls gegen Newton) behauptet und diskutiert. Die Physiker des 19. Jahrhunderts entzogen sich aber meist der Härte des Problems durch die Annahme einer speziellen im Raum ruhenden Substanz, des Lichtäthers. Deshalb hat erst der Michelson-Versuch bzw. Einsteins Deutung dieses Versuchs das philosophische Problem, nun an Hand der Lorentzgruppe, unausweichlich gemacht. Erst bei Einstein wurde das spezielle Relativitätsprinzip aus einer faktisch für gewisse Phänomene gültigen Regel zu einem unentbehrlichen Bestandteil der gewählten Beschreibung von Raum und Zeit. Einstein durfte mit Recht annehmen, damit der Intention der genannten Philosophen (vor allem von Mach und vielleicht Leibniz; Kants Ansichten über Relativität der Bewegung hat er offensichtlich nicht gekannt) erst eine präzise physikalische Gestalt gegeben zu haben.

Diese Gestalt enthält nun aber ein philosophisch
beunruhigendes Problem, auf das Einstein alsbald nach der
Aufstellung der SRth aufmerksam wurde. Das spezielle Relativitätsprinzip leugnet die Existenz eines absoluten Raumes,
ohne doch die Annahme einer allgemeinen Relativität von
Bewegungen zu rechtfertigen. Es steht als empirisch gerecht-
fertigte unbequeme Annahme zwischen zwei scheinbar bequemeren, aber ungerechtfertigten. Die Abgrenzung nach beiden
Seiten sei getrennt diskutiert.

Die Nichtexistenz des absoluten Raumes können wir so
ausdrücken: die Identität eines Raumpunktes im Lauf der
Zeit lässt sich nicht in semantisch konsistenter Weise
behaupten. Zeige ich zweimal nacheinander auf einen Punkt,
so kann ich nicht wissen, ob ich beidemale auf denselben
Punkt gezeigt habe. Ich könnte versuchen, die Identität
des Punktes zu objektivieren, indem ich eine Marke, etwa
eine wiedererkennbare Stelle eines Körpers (kurz ausge-
drückt, einen Körper) in ihm anbringe. Aber die Gruppe,
der gegenüber die Bewegungsgesetze invariant sind, transformiert eine Zustandsbeschreibung, nach welcher der Körper in dem Punkt ruht, in eine solche, in welcher der
Körper mit konstanter Geschwindigkeit auf einer geraden
Bahn läuft, die den Punkt nur in einem bestimmten Zeitpunkt
passiert.

Diese Überlegung konnte schon an Hand der klassischen
Mechanik angestellt werden. Dies tat z.B. L. Lange in der
Gestalt der Einführung der zueinander äquivalenten Iner-
tialsysteme. Einstein hat hinzugefügt, dass auch Zeitpunkte
keine vom Messgerät (der realen Uhr) unabhängige Identi-
tät haben. Seine Überlegung wirkte wie das plötzliche
Aufgehen eines Lichts, weil sie die erste konsequente
Überprüfung dessen war, was hier semantische Konsistenz
genannt wird. Es sei deshalb erlaubt, ihren methodischen
Gehalt (ähnlich wie in Abschnitt 6 für Bohr) in einer fal-
schen und einer richtigen Interpretation zu paraphrasieren.
Gemeinsam ist der Ausgangspunkt: die Wellentheorie des
Lichts führt zum Postulat der Konstanz der Lichtgeschwin-
digkeit, der Michelson-Versuch zum Relativitätspostulat
auch für Licht (und damit für alle bekannten Naturphänome-
ne); beide zusammen zur Lorentz-Invarianz der Naturgesetze.
Nun geht es falsch weiter: "Also kann man absolute Gleichzeitigkeit entfernter Ereignisse nicht mit Uhren feststellen. Was man nicht feststellen kann, existiert nicht. Somit existiert die absolute Gleichzeitigkeit nicht". Richtig ist: "Der Begriff der absoluten Gleichzeitigkeit ist

nicht lorentzinvariant. Wenn alle Naturgesetze lorentzinvariant sind, kann es also keine absolute Gleichzeitigkeit geben. Unser Vorverständnis ist entsprechend zu
korrigieren. Nun könnte jemand einwenden, absolute Gleichzeitigkeit lasse sich doch sogar messen. Demgegenüber
zeigt sich die Konsistenz der Theorie darin, dass Uhren,
die lorentzinvarianten Gesetzen genügen, auch nicht fähig
sind, absolute Gleichzeitigkeit zu messen." Logisch ge-
wendet: "Was gemessen werden kann, existiert" wird als
wahr vorausgesetzt. Die falsche Fassung benutzt die logisch nicht folgende Umkehrung "was nicht gemessen werden
kann, existiert nicht". Die richtige Fassung bestätigt nur
die korrekte Kontraposition: "was nicht existiert, kann
auch nicht gemessen werden." Es sei bemerkt, dass genau
dasselbe Missverständnis bei Kritikern von Heisenbergs
Unbestimmtheitsrelation vorkommt.

Die Physiker, welche die konsequente Deutung der
Quantentheorie aufgebaut haben, also Bohr, Heisenberg und
ihre Nachfolger, haben diese Überlegungen Einsteins stets
als die erste Einführung des Beobachters in eine Diskussion des Sinns physikalischer Begriffe aufgefasst. Einstein
hat sich dagegen verwahrt, von seinem Standpunkt aus mit
Recht, aus zwei Gründen. Erstens kann man bei ihm die
Messung von Längen und Zeitspannen stets als die blosse
Ablesung bestimmter Zustände von Maßstäben und Uhren ange-
sehen werden, die auch dann vorliegen, wenn niemand sie
beobachtet. Wieweit dieses Argument auf die quantentheoretische Messtheorie übertragbar ist, bleibe hier uner-
örtert; das "Paradox" von Einstein-Rosen-Podolsky soll
zeigen, dass sie es nicht ist. Zweitens aber sind zwar
weder Raum noch Zeit je für sich im hier definierten Sinne
absolut, wohl aber das vierdimensionale Raum-Zeit-Kontinuum, Minkowskis "Welt". Ein Weltpunkt, oder, wie Einstein
gern sagte, ein "Ereignis" wird in der speziellen und in
der allgemeinen Relativitätstheorie als objektiv identifiziert behandelt[3]. Dies wird freilich nicht mehr im
Sinne semantischer Konsistenz begründet, sondern als quasi
evident vorausgesetzt. Die Mannigfaltigkeit von Raumpunk-
ten und ebenso die Mannigfaltigkeit von Zeitpunkten eines
gewählten Inertialsystems erscheint unter diesem Aspekt
als ein konventionelles Ordnungsschema in einer nicht
konventionellen Ereignismannigfaltigkeit. Die semantische
Inkonsistenz dieser Annahme einer objektiven Ereignis-
menge wird erst in der Quantentheorie zum Thema.

Für die Begründung der physikalischen Geometrie

mindestens so wichtig wie die Nichtobjektivität absoluter
Geschwindigkeiten ist aber die Objektivität absoluter
Beschleunigungen in der klassischen Mechanik und der SRth.
Die Galilei-Transformation drückt, wie Newton klar sah und
durch den Eimerversuch nachwies, nicht eine allgemeine
Relativität der Bewegung aus, sondern eine Folge eines
speziellen dynamischen Gesetzes, des Trägheitsgesetzes.
Die Tatsache, dass die Newtonsche Bewegungsgleichung
von zweiter Ordnung in der Zeitableitung ist, hat zur
Folge, dass nur Geschwindigkeiten, nicht aber Beschleunigungen als relativ aufgefasst werden dürfen. Diese
schlechterdings nichttriviale Tatsache wird man heute wohl
am liebsten damit in Zusammenhang bringen, dass die Newtonsche Gleichung die Eulersche Gleichung eines euklidisch
invarianten Variationsprinzips ist. Dieses seinerseits
kann man im Sinne von Dirac[4] als Huygenssches Prinzip
der Schrödingerwellen auffassen. Auch in diesem Punkte
finden also die Grundfakten der klassischen Physik ihre
nächste Erklärung in der Quantentheorie (vgl. dazu Abschnitt 10). Die Ungeklärtheit dieses Problems war für
Einstein der Anlass zur Suche nach einer allgemeinen Relativitätstheorie. Ehe wir ihm hierin folgen, sei aber noch
die Beziehung der hier besprochenen Theorien zur Erfahrung knapp methodologisch charakterisiert.

Alle diese Theorien gehen von empirisch bewährten
Gesetzmässigkeiten aus, die von ihrem Vorverständnis
her keineswegs selbstverständlich sind, sich aber gegen
empirische Falsifikationsversuche in einer für die scientific community überzeugenden Weise als resistent erwiesen
haben. Man kann diese Gesetzmässigkeiten den harten Kern
der betr. Theorien nennen. Sie werden dann als schlecht-
hin allgemeingültige Prinzipien hypothetisch postuliert.
Wer dieses Postulat akzeptiert, modifiziert damit sein
Vorverständnis. Vom neuen Vorverständnis aus sind die
ursprünglich nichttrivialen empirischen Grundfakten notwendige, keiner weiteren Erklärung bedürftige Phänomene.

Der harte Kern des "Galileischen" Relativitätsprin-
zips der klassischen Mechanik ist das zunächst empirische
Faktum des Trägheitsgesetzes. Postuliert man das Relativitätsprinzip als Naturgesetz, dann haben nicht Raumpunkte, wohl aber Trägheitsbahnen objektive Realität. Dann
ist das Trägheitsgesetz eine selbstverständliche Konsequenz des postulierten Naturprinzips der Relativität.

Der harte Kern der SRth ist das zunächst empirische
Faktum des negativen Ausfalls des Michelson-Versuchs.

Fordert man Einsteins zwei Postulate als Naturgesetze,
dann hat die absolute Geschwindigkeit des Michelson-
Apparates keine objektive Realität. Dann ist Michelsons
Ergebnis die selbstverständliche Konsequenz der postulierten Prinzipien.

9. ALLGEMEINE RELATIVITÄTSTHEORIE

Der harte Kern der ARth ist das zunächst empirische Faktum
der Gleichheit der schweren und trägen Masse. Man kann
deshalb postulieren, dass es eine besondere Kraft "Gravitation" gar nicht gibt, sondern stattdessen eine Riemannsche Geometrie des Raum-Zeit-Kontinuums. Dann ist die
Gleichheit beider Massen eine selbstverständliche Konsequenz.

Historisch fand Einstein den Weg zu diesem Ergebnis
über die Suche nach einem allgemeinen Relativitätsprinzip.
Dabei gingen aber einige der für seinen ursprünglichen
Ansatz charakteristischen Züge verloren, und die Debatte
über den Sinn des Begriffs der allgemeinen Relativität
hat sich bis heute nicht ganz befriedigend aufgelöst.

Machs Motiv für die Forderung allgemeiner Relativität
war die ontologische Annahme, dass nur Körper, nicht aber
Raumpunkte objektive Realität haben; deshalb sollten die
Kräfte zwischen Körpern nur von ihren Relativkoordinaten,
nicht aber von ihrem Verhältnis zu einem "absoluten Raum"
bestimmt sein.In dieser Form ist die Forderung mit der
Einführung von Feldttheorien hinfällig, zumal einer Feld-
theorie der Raummetrik. Einstein schreibt dem Raum nicht
wie Mach keine, sondern mehr physikalische Eigenschaften
zu als Newton. Ob die Forderung, die Einstein dann "Mach-
sches Prinzip" genannt hat, physikalisch berechtigt ist,
ob also das metrische Feld durch die Materieverteilung
vollständig determiniert sein soll, bleibt solange frag-
lich, als keine physikalische Theorie der Materie vorliegt.

In der Gestalt, die die Theorie unter Einsteins
Händen erreicht hat, hat jeder Weltpunkt bestimmte physikalische Eigenschaften, nämlich die Werte des metrischen
Tensors und des Materietensors. Die Erfüllung der Forderung nach allgemeiner Relativität fand Einstein nun
darin, dass zur Bezeichnung dieser Punkte völlig beliebige Koordinaten gewählt werden dürfen und dann die
Gesetze der Theorie gegen beliebige Transformationen dieser
Koordinaten invariant sein sollen (allgemeine Kovarianz).

Damit ist aber die semantische Konsistenz, die Machs und der SRth Ziel war, bis auf weiteres preisgegeben. Mach wollte die Theorie dem Vorverständnis anpassen: nur Beziehungen zwischen Körpern sollten in ihr vorkommen. Einstein hat den Feldbegriff auch für die Geometrie eingeführt und muss das Vorverständnis entsprechend modifizieren. Die ARth ist damit ein neues schönes Beispiel "antihierarchischen" Denkens. Aber ihre semantische Konsistenz würde erfordern, dass man erfährt, wie die jeweiligen Koordinatenwerte empirisch bestimmt werden, also wie sich Massstäbe und Uhren nunmehr verhalten. Gerade die Beliebigkeit der Koordinatenwahl nimmt aber auf dieses Problem gar keinen Bezug. Die Theorie braucht nicht semantisch inkonsistent zu sein, aber ihre semantische Konsistenz ist kein konstruktives Element ihres Aufbaus. In der Tat müssen diese Fragen ebenfalls dem ungelösten Problem einer Theorie der Materie zugeschoben werden.

Das hat jedoch zur Folge, dass der physikalische Sinn der allgemeinen Kovarianz undeutlich bleibt. Wie Kretschmann zuerst hervorgehoben hat, kann man durch explizite Einführung der Komponenten des metrischen Tensors in die Feldgleichungen diese stets allgemein kovariant schreiben. Demnach ist die allgemeine Kovarianz überhaupt keine Forderung an den Inhalt der Naturgesetze, sondern nur an deren Schreibweise. Einstein hielt aber daran fest, es sei eine sachhaltige Forderung, die Naturgesetze sollten in der allgemein kovarianten Schreibweise einfach sein. In der Tat sind seine Feldgleichungen (mit kosmologischem Glied) die einzigen allgemein kovarianten, die keine höheren als zweite Ableitungen der g_{ik} enthalten. Das Prinzip der Einfachheit der Naturgesetze hat eher ästhetischen als begrifflichen Charakter. Vermutlich appelliert es an eine Gestaltwahrnehmung guter Physiker für noch unverstandene, aber im Prinzip angebbare begriffliche Strukturen.

Zwei Fragen können auch heute aufgeworfen werden: Was bedeutet die Lorentz- (Poincaré-)Gruppe in der ARth? und : Was bedeutet die Beschränkung auf stetige topologische Transformationen?

Die Poincaré-Gruppe ist die Invarianzgruppe der ARth im Kleinen, also lokal in linearer Näherung. In der klassischen Mechanik entsprechen den 10 Parametern dieser (bzw. der Galilei-) Gruppe die 10 allgemeinen Integrale der Bewegungsgleichungen. Nun hat in der Punktmechanik das n-Körper-Problem 6 n Integrale. Die offensichtlichen

Symmetrien, die den 10 sog. allgemeinen Integralen entsprechen, sind durch das Vorverständnis der Punktmechanik
ausgezeichnet, nach dem die 6 n verallgemeinerten Koordinaten eine Darstellung zulassen müssen als Koordinaten
und Impulse von n Massenpunkten, deren jeder an den Symmetrien von Raum und Zeit gemäss der Galilei (oder Poincaré-)
-Transformation teilhat. Dieses Vorverständnis kann natürlich wiederum nur weiter erklärt werden durch eine Theorie
der Materie, die angibt, warum es Körper gibt, die als
Massenpunkt approximiert werden können. Analog kann man für
die 10 Integrale der von Einstein benutzten formalen Kon-
tinuumsdarstellung der Materie argumentieren, z.B. durch
Grenzübergang von der Punktmechanik her. Diese Betrach-
tungsweise entspricht nun genau der lokalen linearen
Approximation in der ARth. Die nichtlinearen Glieder heben
die Poincaré-Symmetrie auf und damit das Mittel der Unterscheidung der 10 allgemeinen Integrale von den übrigen. Der
allgemeinen Kovarianz entsprechen dann die Bianchi-Identi-
täten.

Zur Topologie: in einem hierarchischen Aufbau der
Geometrie nach dem Erlanger Programm beginnt man mit
beliebigen Punkttransformationen, die man dann schrittweise einschränkt durch die Forderung, dass gewisse Beziehungen zwischen Punkten invariant bleiben sollen, zuerst
die Topologie, dann lineare (projektive oder affine),
schliesslich metrische Beziehungen. Eine umgekehrte Reihenfolge würde sich nahelegen, wenn man Beziehungen zwischen
möglichst wenigen Punkten zugrundelegt. Eine metrische
Beziehung (Abstand) besteht zwischen zwei Punkten, eine
lineare (auf einer Geraden, Ebene, ... liegen) zwischen
wenigstens drei Punkten, eine topologische (Häufungspunkt
sein etc.) zwischen unendlich vielen Punkten. Nun definiert in der Tat eine Metrik auch eine Topologie. In einer
semantisch konsistenten Physik erscheint es plausibel,
dass räumliche und zeitliche Abstände gemessen werden
können. Also sollte man das Raum-Zeit-Kontinuum wohl
nicht primär als topologischen Raum von "Ereignissen"
auffassen, dem dann eine Metrik aufgeprägt wird, sondern
primär durch metrische Relationen bestimmen, welche bei
fingierter absoluter Messgenauigkeit, auch eine Topologie
festlegen. Die Metrik muss dabei nicht die pseudoeukli-
dische der Minkowskiwelt sein, für welche Punkte mit
lichtartigem Abstand den metrischen Abstand Null haben,
sondern eine durch die positiv definite Summe von räum-
lichem und zeitlichem Abstand definierte Metrik. Zwar ist

diese Metrik selbst nicht lorentzinvariant, wohl aber die
durch sie definierte Topologie. Damit hat man dann die
Freiheit, die quantentheoretische Begrenzung des Raumbe-
griffs einzuführen.

10. QUANTENTHEORIE DES RAUMS

In der nichtrelativistischen Quantentheorie ist das Problem
der semantischen Konsistenz der Geometrie völlig beiseite-
gelassen. Nach Helmholtz, Poincaré, Dingler und Einstein
muss dies wie ein Rückfall in eine primitive Unreflektiertheit wirken. Das eigentlich erklärungsbedürftige Phänomen
aber dürfte der ausserordentliche empirische Erfolg einer
geometrisch so unreflektierten Theorie sein. Diese Behauptung sei hier zunächst erläutert.

Das Vorverständnis der euklidischen Geometrie lässt
sich am besten mit der Existenz frei verschieblicher fester
Körper formulieren. Die physische Existenz solcher starrer
Körper (in der Näherung, in der sie faktisch existieren)
erklärt die Quantentheorie. Sie hat damit erst voll ins
Bewusstsein der Physiker gehoben, eine wie schwierige, mit
der klassischen Kontinuumsdynamik wohl unlösbare Aufgabe
diese Erklärung ist. Unter den Philosophen war sich vielleicht Kant der Schwierigkeit dieser Aufgabe am klarsten
bewusst (Metaphysische Anfangsgründe der Naturwissenschaft,
2. Hauptstück, und Opus Postumum). Die Physiker stiessen
darauf in der statistischen Thermodynamik (Boltzmann).
Ohne Atome, die keine inneren Freiheitsgrade mehr haben,
gibt es kein thermodynamisches Gleichgewicht. Also darf man
die Dynamik des Kontinuums nicht auf das Innere der Atome
anwenden. Punktmechanische Atommodelle wie das von Ruther-
ford scheitern in der klassischen Physik an der Kontinuumsdynamik der sie zusammenhaltenden Kräfte, wie Bohr erkannte.
Bohr und nach ihm die Quantenmechanik erklärte die Existenz
formkonstanter Atome durch einen radikalen Bruch mit der
klassischen Dynamik. Vom Standpunkt des bewusstgemachten
Vorverständnisses der Geometrie aus ist es dann aber
mindestens sehr verblüffend, dass die Quantenmechanik
auch auf das Innere des Atoms, bis zu beliebig kleinen
Längen hinunter (empirisch gerechtfertigt jedenfalls bis
zu 10^{-12}cm, also einem Zehntausendstel des Atomradius)
euklidische Geometrie anwendet, obwohl es offenbar unmöglich ist, diese kleinen Räume mit starren Körpern,
welche aus "Subatomen" bestehen müssten, auszumessen,
Charakteristischerweise bedienen sich die quantentheore-

tischen Gedankenexperimente zur Ortsmessung nicht der starren Maßstäbe, sondern des Mikroskops; die übliche Theorie
des Mikroskops aber setzt bereits die euklidische Geometrie
voraus.

Es ist kein Wunder, dass eine geometrisch so undurch-
dachte Theorie in Schwierigkeiten kommt, wenn sie mit der
höchsten bisher erreichten Reflexionsstufe bezüglich semantischer Konsistenz der Geometrie vereinigt werden soll,
nämlich der speziellen Relativitätstheorie. Formal befriedigend geglückt ist diese Vereinigung bisher nur für kräf-
tefreie Theorien, also gerade unter Vernachlässigung des
Zugs der Wirklichkeit, an dem die semantische Konsistenz
(die Theorie der Messung) hängt, der Dynamik. Die Dynamik
scheint eine Einschränkung der realen Anwendbarkeit der
Geometrie bei sehr kleinen Längen zur Folge zu haben, für
die es heute aber noch keine allgemein anerkannte Beschreibung gibt. Hingegen gibt es keine Indizien dafür, dass
diese Einschränkung zugleich eine Einschränkung des Geltungsbereichs der Quantentheorie wäre. Nun bedarf ein
axiomatisch konsequenter Aufbau der Quantentheorie (Jauch
1968, Piron, Drieschner[5]) in der Tat des Ortsbegriffs nicht.
Man kann also versuchen, zunächst eine abstrakte Quantentheorie zu errichten, und dann in ihr einen Parameter oder
gar Operator zu finden, der mit dem Ort identifiziert werden könnte. Diese Identifikation müsste dann sowohl den
Erfolg der euklidischen Geometrie unterhalb des Bereichs
erklären, in dem es feste Körper gibt, wie auch die Einschränkung für noch kleinere Längen.

Nun lässt die abstrakte Quantentheorie zunächst eine
ganz ausserordentlich umfassende Transformationsgruppe zu,
nämlich die unitäre Gruppe des ganzen Hilbertraums. Die
Einschränkung des hierin implizierten Konventionalismus
geschieht durch die Wahl der Dynamik. Die physikalische
Geometrie ist auch in der Quantentheorie durch diejenige
Gruppe bestimmt, welche die Dynamik invariant lässt.
Realiter wird in der heutigen Quantenfeldtheorie der Aufbau
freilich umgekehrt vollzogen. Man kennt vorweg die geometrisch relevante Untergruppe der dynamischen Gruppe, nämlich die Poincaré-Gruppe, und wählt die Dynamiken unter
denjenigen aus, die ihr gegenüber invariant sind. Dieses
Verfahren bedeutet aber, dass man über die geometrische
Unreflektiertheit der unrelativistischen Quantenmechanik
nicht hinausgekommen ist. Man hat lediglich die euklidisch-
galileische Geometrie durch die Minkowskische ersetzt,
auch für den atomaren und subtomaren Bereich, in dem

Einsteins Rechtfertigung der semantischen Konsistenz dieser
Theorie nicht in manifester Weise anwendbar bleibt.

Daher schlage ich ein fundamental anderes Verfahren
vor. Die abstrakten Eigenschaften einer Quantentheorie
der Wechselwirkung sollen vorweg studiert werden, und es
soll versucht werden, aus ihnen die dynamische Gruppe
und daraus die Geometrie herzuleiten.

In der abstrakten Axiomatik wird die Quantentheorie
normalerweise als die Theorie der Zustände eines einzelnen
Objekts, oder, wie man auch sagt, Systems formuliert. Der
Anschluss dieser abstrakten Theorie an die Wirklichkeit
geschieht über den Begriff der Wahrscheinlichkeit. Eine
Analyse des Vorverständnisses dieses Begriffs lässt schwer-
lich eine andere Deutung zu als dass er eine Prognose
relativer Häufigkeiten beobachtbarer Ereignisse bezeichnet. In dieser Fassung tritt das beobachtende und prog-
nostizierende Subjekt in die Theorie ein, zwar nicht als
Gegenstand ihrer Beschreibung, aber als Voraussetzung ihres
Sinns. Nun ist das Subjekt nicht nur Beobachter, sondern
zugleich Teil der Welt. Wir sind, wie Bohr zu sagen pfleg-
te, an die alte Tatsache erinnert worden, dass wir zugleich
Zuschauer und Mitspieler im Schauspiel des Daseins sind.
Eine semantisch konsistente Quantentheorie müsste diesen
Sachverhalt beschreiben. Der erste Anlauf dazu ist die
Messtheorie. Sie beschreibt die Messapparate als besondere
quantentheoretische Objekte und lässt erst die Art der
Kenntnisnahme von den Messapparaten durch das Subjekt
undiskutiert . Sie rechtfertigt diesen Schritt durch den
in hinreichender Näherung klassischen Charakter der Messapparate (Bohr), den sie dann freilich möglichst explizit
nachweisen muss (Jauch 1968). Diese Seite des Problems
möge hier undiskutiert bleiben. Wir studieren jetzt nur
die Folgen der Tatsache, dass eine semantisch konsistente
Quantentheorie notwendig eine Theorie nicht eines einzigen,
sondern mehrerer wechselwirkender Objekte sein muss.

Vielleicht erscheint es sehr umständlich, die Vielheit der Objekte und ihre Wechselwirkung durch die Forderung semantischer Konsistenz zu begründen, da doch jeder
Physiker weiss, dass es viele wechselwirkende Objekte
gibt. Es wäre aber sonst nicht selbstverständlich, dass
dieses Moment im Vorverständnis der Physiker den Übergang
von der klassischen zur Quantentheorie überleben muss.
Die Behandlung einer Vielheit von Objekten als ein einziges
Objekt enthält ja einen besonders krassen Unterschied der
Quantentheorie von der klassischen Physik und Ontologie.

Auch im klassischen Denken kann man eine Vielheit von
Objekten als ein Gesamtobjekt beschreiben; jeder wohldefinierte Zustand des Gesamtobjekts legt dann zugleich
die Zustände seiner Teilobjekte fest. In der Quantentheorie
hingegen bilden unter den Zuständen des Gesamtobjekts
diejenigen, in denen die Teilobjekte wohldefinierte Zustände haben (also im eigentlichen Sinn als Teilobjekte existieren) eine Teilmenge vom Mass Null. Eben darum erlegt die
semantische Konsistenz der Quantentheorie eine zusätz-
liche Forderung auf.

Wir versuchen diese Forderung so zu formulieren: In
der Näherung, in der die Vorgänge beobachtbar sind, soll
die Identität der die Beobachtung ermöglichenden Objekte
über eine hinreichende Zeitspanne gewahrt bleiben. Dadurch
wird die Gruppe der zulässigen Transformationen einge-
schränkt. Sie müssen jedes dieser Objekte in sich transformieren. Diese Forderung lässt sich weiter speziali-
sieren, wenn wir vom Begriff gleichartiger Objekte Gebrauch
machen. Als gleichartig sollen Objekte gelten, deren
Zustandsmannigfaltigkeiten und dynamische Gesetze iso-
morph aufeinander abgebildet werden können. Ein isoliertes
Objekt definiert eine Transformationsgruppe, die seine
innere Dynamik invariant lässt. Besteht ein Gesamtobjekt
aus vielen gleichartigen Objekten, so wollen wir fordern,
dass seine Dynamik (die eine Wechselwirkung seiner Teilobjekte enthält) invariant sei gegen eine Transformation,
in der jedes der Teilobjekte gleichzeitig durch dasselbe
Element der Transformationsgruppe der isolierten Teilobjekte transformiert wird. Gewöhnlich nennt man das: die
Wechselwirkung soll dieselbe Symmetriegruppe haben wie
die freie Bewegung. Es sei hier nicht versucht, diese
Fassung der Forderung über die vorgebrachten Plausibili-
tätsargumente hinaus streng zu begründen. Sie steht im
Einklang mit den heutzutage üblichen Annahmen.

Die Frage ist nun, in wie kleine Teilobjekte man ein
empirisch gegebenes Objekt zerlegen kann. Der Begriff
des Teilchens, also auch des Elementarteilchens, setzt
die Beschreibung im Ortsraum schon voraus, die wir erst
begründen wollen. Auch ist die heutige Elementarteilchen-
physik eine Physik der Zerlegung und Erzeugung sogenann-
ter Elementarteilchen. Nach klassischer philosophischer
Auffassung sollte ein Atom ein Gebilde sein, das unteil-
bar ist, weil es begrifflich keine Teile hat, nicht eines,
dessen denkbare Teile empirisch bisher nicht getrennt
werden können. Ausgedehnte "Atome" haben diesem Ideal nie

entsprochen, auch Massenpunkte haben noch immer eine additive Eigenschaft, eben die Masse. In der Quantentheorie kann man jeden Hilbertraum multiplikativ auf zweidimensionale Teilräume zurückführen, wobei auch Dimensionszahlen, die keine reinen Zweierpotenzen sind, z.B. durch Symmetrievorschriften, erzeugbar sind. Die Observablen eines zweidimensionalen Zustandsraumes definieren einfache Alternativen. Sie bedeuten die begrifflich ärmsten möglichen empirischen Entscheidungen. Objekte, die einen solchen zweidimensionalen Zustandsraum haben, sind also in der Quantentheorie die einzigen Kandidaten für einen philosophisch strengen Begriff der Unteilbarkeit. Wir machen die Hypothese, dass alle Objekte aus solchen "Urobjekten" (kurz, um die Abstraktion kalauerhaft auszudrücken, "Uren") bestehen. Die Theorie der Ure soll anderswo im Zusammenhang dargestellt werden (vgl. 5) 6)). Hier seien nur ihre Grundzüge angedeutet.

Die Hypothese über die Symmetrie der Dynamik besagt, dass die Dynamik jedes aus Uren bestehenden Gesamtobjekts invariant sein muss gegen die Symmetriegruppe des isolierten Urs. Diese besteht (ausser der zunächst beiseitegelassenen Multiplikation des Zustandsvektors mit einem konstanten Phasenfaktor) aus der SU(2). Der Hilbertraum jedes Gesamtobjekts kann natürlich entsprechend seiner Zerlegung in Ure als Darstellungsraum der SU(2) aufgefasst werden. Nun ist es formal zweckmässig, die Vektoren eines Darstellungsraumes einer Gruppe als Funktionen in einem homogenen Raum der Gruppe zu schreiben. Der grösste homogene Raum einer Gruppe ist die Gruppe selbst. Also wird es zweckmässig sein, die Zustandsvektoren aller Objekte als Funktionen auf der SU(2) zu schreiben. Die SU(2) ist, als Raum betrachtet, ein reeller dreidimensionaler sphärischer Raum. In diesem Raum sind die Transformationen der SU(2) selbst Cliffordsche Schiebungen eines festen Schraubungssinnes, sagen wir Rechtsschrauben. Die Dynamik aller Objekte muss gegen diese Schiebungen invariant sein. Machen wir die Zusatzhypothese, sie sei auch gegen die Linksschrauben invariant, also die volle metrische Gruppe unseres sphärischen Raums; so definiert diese Invarianzgruppe gemäss Kleins Erlanger Programm eine physikalische Geometrie, d.h., sie definiert eben diesen sphärischen Raum als den Ortsraum der Physik. Ohne die Zusatzhypothese erhielte man eine Dynamik, die im selben Raum darstellbar wäre, aber unter Paritätsverletzung.

Der Vorschlag ist, diese Herleitung ernst zu nehmen,
also in ihr eine Deduktion der physikalischen Geometrie
aus einer semantisch konsistenten Quantentheorie einfacher
empirischer Alternativen zu sehen. Zur Überprüfung muss
zunächst die Beschreibung der Zeit studiert und womöglich
festgelegt werden. Wenn dies eindeutig gelänge, müsste
damit die Basis sowohl für die Theorie der Elementarteilchen wie der Kosmologie (Theorie des Weltraums) gegeben
sein, und die Ausführung müsste den Vergleich mit der Erfahrung ermöglichen. Die Probleme der Geometrie werden
dabei in drei sukzessiven Modellen oder Annäherungsschrit-
ten auftreten.

Betrachtet man, als einfachstes Modell, die ganze Welt als
ein Objekt, das aus einer zeitlich konstanten endlichen
Anzahl von Uren besteht, so ist unser sphärischer Raum
der Weltraum, und das Modell entspricht dem statischen
Einstein-Kosmos. Die Impulse aller Objekte (der Teilobjekte der Welt) sind gequantelt. Das isolierte Ur ist das
Objekt mit minimalem Impuls. Alle Operatoren haben diskrete
Spektren. Also gibt es keinen kontinuierlichen Ortsopera-
tor. Der kontinuierliche Ort im Weltraum ist ein Parameter.
Die Metrik im Weltraum ergibt sich aus der Hilbertraummetrik des Urs. Der physikalische Sinn der Hilbertraummetrik
liegt im Wahrscheinlichkeitsbegriff. Als Abstand zweier
Zustände des Urs kann die Wahrscheinlichkeit gelten, den
einen nicht zu finden, wenn der andere vorliegt. Die Abstände werden also nicht am Einzelur, sondern als relative Häufigkeiten an grossen Gesamtheiten von Uren gemessen. Der Ort als beobachtbare Grösse ist ein klassischer
Grenzfall. Je mehr Ure in einem Teilobjekt, desto höher
sein möglicher Impuls, desto genauer ist es also lokali-
sierbar. Die Äquivalenz aller Punkte, die bisher als unbe-
gründetes Postulat erschien, ist nun eine Folge der Wahl
eines homogenen Raums der Gruppe als Grundlage unserer
Darstellung. Sie ist in diesem Sinne konventionell, aber
eben die einem Vorverständnis sich am ehesten aufdrängende
(in ihm unbewusste) Konvention. Die Wechselwirkung muss
zur Bildung von Teilchen führen. Sei ν die Anzahl von
Uren pro Elementarteilchen, n die Anzahl der Elementarteilchen in der Welt, N die Anzahl der Ure in der Welt,
so ist $N=n\cdot\nu$ .Die Anzahl ν begrenzt die Lokalisierbarkeit
eines Elementarteilchens. Nennt man die verbleibende Unge-
nauigkeit seiner Lokalisierung seinen Radius r und den
Weltradius R, so wird vermutlich gelten $R=r\cdot\nu$. R und r
sind die beiden "natürlichen" Längeneinheiten für

kosmologische bzw. atomare Längenmessung.

Dieses Modell war nichtrelativistisch. Das ist natürlich, denn die zugrundegelegte abstrakte Quantentheorie
benutzt den Raumbegriff nicht, wohl aber den Zeitbegriff.
Sie benutzt einen absoluten Zeitparameter, ohne sich um die
semantische Konsistenz des Zeitbegriffs, also die Zeitmessung, zu kümmern. Es gibt für sie eine absolute Gleichzeitigkeit aller Zustände eines Objekts. Will man die spezielle Relativitätstheorie berücksichtigen, so muss man die
Axiomatik der Quantentheorie anders aufbauen. Formal ist
das möglich, indem man die Anzahl der Ure zeitlich variabel
oder unendlich sein lässt und die Zerlegung eines Objekts
in Ure verschieden vornimmt je nach der Wahl der Zeitkoor-
dinate, d.h. des die Zeit messenden Subjekts oder seiner
Uhr. Semantisch bedeutet dies, dass die fundamentalen
Alternativen nicht lorentzinvariant definiert sind. Z.B.
sind zwei Spinrichtungen, die in einem Lorentzsystem ent-
gegengesetzt sind, dies in einem anderen nicht. Die Zeitmessung kann nun genau wie die Ortsmessung als eine statistische Messung an vielen Uren definiert werden. Dieses
Modell ist bisher nicht durchgeführt.

Ein drittes Modell müsste die allgemeine Relativitätstheorie einbeziehen. Im ersten Modell haben wir schon einen
konstant gekrümmten Raum. Sein Krümmungsmass ist $1/R$ oder,
wenn wir verabreden, die atomare Längeneinheit $r = 1$ zu
setzen, ist es $1/\nu$. Mit der üblichen Schätzung $R=10^{40}r$
ist dies gerade die Krümmung, die man für einen Einstein-
kosmos annehmen müsste. D.h. <u>wenn</u> wir das Problem lösen
können, die Wechselwirkung der Ure so zu bestimmen, dass
sie die richtige Anzahl $\nu = 10^{40}$ von Uren pro Elementarteilchen liefert, dann werden wir automatisch für die
Krümmung eines Einstein-Kosmos den richtigen Wert erhalten,
wenn wir den empirischen Wert der Einsteinschen Gravita-
tionskonstanten in Einsteins Grundgleichung einsetzen. Gäbe
es nur die mittlere kosmische Raumkrümmung, so wäre dies
die vollständige Theorie der Raumkrümmung, d.h. nach
Einstein, der Gravitation. Nun gibt es aber vor allem das
lokal variierende Gravitations-, also, nach Einstein, metrische Feld. Für dieses lässt sich nur sagen: Wenn es einen
Anteil der Wechselwirkung gibt, der sich als Raumkrümmung
schreiben lässt, so muss er in der Näherung, in der
höchstens zweite Ableitungen der Feldgrössen vorkommen, aus
Invarianzgründen der Einsteinschen Gleichung genügen. Ort
und Zeit sind aber in dieser Theorie nur in einem relativ
zu den Uren klassischen Grenzfall definiert. Also ist es

nicht a priori klar, ob eine Quantelung der Gravitation
sinnvoll ist, und ebensowenig, ob es einen Grund gibt,
die höheren Ableitungen aus der Grundgleichung fortzulassen.
   Wenn man überhaupt wagt, eine wegen ihrer Schwierigkeit bisher nicht durchgeführte Theorie in dieser Weise zu
schildern, so ist der Sinn davon nur, durch ihre Denkmöglichkeit darauf hinzuweisen, welche Probleme notwendigerweise auftauchen werden, wenn man versucht, Quantentheorie
und Relativitätstheorie semantisch konsistent zu vereinigen.

11. ERKENNTNISTHEORETISCHE REFLEXION

Zum Abschluss sei versucht, die etwas gewundenen Gedanken-
gänge dieses Aufsatzes einmal in der ursprünglichen Reihenfolge knapp zu resumieren, und dann in entgegengesetzter
Reihenfolge nochmals zu durchlaufen, um die Konsequenzen
der bisherigen Resultate für die Ausgangsfrage zu über-
prüfen.
   1. Axiomatik. Die von den Griechen entwickelte axiomatische Darstellung der Geometrie rechtfertigt nicht die
Meinung, die Axiome seien evident und die Folgesätze müssten
sich daher mit Sicherheit in der Erfahrung bewähren. In
Hilberts Auffassung der Axiomatik entscheidet die Mathematik weder über die Wahrheit noch über den Sinn der Axiome.
Gibt es eine Methode, zu entscheiden, ob bestimmte Axiomensysteme die räumliche Erfahrung richtig beschreiben?
   2. Empirismus. Die Meinung, man könne empirisch über
die Wahrheit geometrischer Axiome entscheiden, scheitert
zunächst an der empirischen Unbestimmtheit der verwendeten
Begriffe. Fände man in einem Lichtstrahlendreieck eine
Winkelsumme ungleich zwei Rechten, so bliebe offen, ob
vielleicht die Lichtstrahlen keine Geraden sind.
   3. Hierarchismus. Man nimmt traditionell an, gewisse
Wissenschaften seien anderen so vorgeordnet, dass Ergeb-
nisse der letzteren die Aussagen der ersteren nicht modifizieren können; so sei die Logik allen Wissenschaften vorgeordnet, und die Mathematik, speziell auch die Geometrie,
der Physik. Dies soll hier bestritten werden. Eben darum
werden zunächst Verteidigungen des Hierarchismus studiert.
   4. Kants Fragestellung. Kant begründet die Mathematik
auf Konstruktion in der reinen Anschauung und ihre hierarchische Vorordnung vor der Physik darauf, dass sie Bedingung der Möglichkeit von Erfahrung ist, da die reine Anschauung zugleich die Form der empirischen Anschauung ist.
Bohrs Auffassung der Messung übernimmt, was hiervon in der

modernen Physik haltbar erscheint. Jedoch ist die empirische
Anschauung weder euklidisch noch nichteuklidisch, sondern
unpräzise, und die reine Anschauung ist schon ein Produkt
des Denkens.

5. Konventionalismus. Mathematisch formulierte Gesetze
der Physik lassen Transformationsgruppen zu. Eine hierbei
nicht invariante Formulierung darf als konventionell betrachtet werden, eine invariante hingegen als echte Aussage.
Welche Transformationen sind zuzulassen? Fasst man den Raum
als Punktmannigfaltigkeit auf, so sind die Punkte nicht
verschieden, jeder sollte also in jeden transformierbar
sein. Punkte sind nur durch Hinzeigen unterscheidbar. Wie
garantiert man aber, dass der Punkt, auf den man zweimal
gezeigt hat, derselbe Punkt war?

6. Operative Begründung der euklidischen Geometrie.
Helmholtz und Dingler begründen die euklidische Geometrie
auf Annahmen über die Möglichkeit bestimmter Operationen
mit starren Körpern. Sie können aber die Möglichkeit nicht
ausschliessen, dass diese Operationen real undurchführbar
werden.

7. Semantische Konsistenz. Auch die Wissenschaftstheorie kann der speziellen Wissenschaft nicht hierarchisch
vorgeordnet werden; sie bedarf der Korrektur durch die
Geschichte der Wissenschaft. In der realen Geschichte geht
eine Theorie stets von einem Vorverständnis aus, mit dessen
Hilfe sie ihren mathematischen Formalismus auf die Wirklichkeit bezieht. In diesem Sinne ist das Vorverständnis der
Theorie vorgeordnet. Aber die Theorie kann das Vorverständnis korrigieren. In diesem Sinne ist die Vorordnung nicht
hierarchisch. Wenn die Theorie ihr Vorverständnis selbst
widerspruchsfrei zu deuten vermag, heisse sie semantisch
konsistent.

8. Spezielle Relativitätstheorie. Ihr Vorverständnis
enthält die euklidische Geometrie und die Newtonsche
Mechanik. Als Formulierung des Vorverständnisses für die
euklidische Geometrie können die Helmholtz-Dinglerschen
Forderungen an starre Körper gelten, für die Newtonsche
Mechanik das Trägheitsgesetz. Die Galilei-Transformation
relativiert die Identität des Raumpunkts, die Lorentz-
Transformation auch die des Zeitpunkts. Einsteins Diskussion von Uhren und Massstäben ist das klassische Modell einer
Herstellung semantischer Konsistenz durch Kritik des Vorverständnisses an Hand der entwickelten Theorie. Methodisch
geschieht dies gemäss Einsteins Diktum: "Erst die Theorie
entscheidet, was messbar ist". Anfangs empirische Fakten

wie Trägheitsgesetz und Michelson-Versuch werden am Ende
notwendige Konsequenzen der postulierten Prinzipien der
Theorie.

9. Allgemeine Relativitätstheorie. Sie verknüpft die
durch die empirische Gleichheit von schwerer und träger
Masse angeregte Reduktion der Gravitation auf Raum- Zeit -
Krümmung mit dem davon logisch unabhängigen Prinzip der
allgemeinen Kovarianz. "Punkte" mit Identität sind die
Ereignisse. Das Kovarianzprinzip ist nur verbunden mit der
Forderung der Einfachheit physikalisch gehaltvoll. Die Forderung semantischer Konsistenz wird preisgegeben bzw. auf
die zukünftige Theorie der Materie abgewälzt. Die lokale
Auszeichnung der Poincaré-Gruppe lässt sich auf die in
linearer Näherung mögliche Zerlegung der Materie in unabhängige Teile (z.B. Massenpunkte) zurückführen. Die Auszeichnung topologischer Transformationen dürfte auf der
Raum- und Zeitmetrik beruhen.

10. Quantentheorie des Raums. Das Vorverständnis der
metrischen Geometrie setzt Annahmen über formkonstante
Körper voraus, die erst die Quantentheorie begründet. Die
unrelativistische Quantentheorie aber setzt die euklidische Geometrie auch in Dimensionen voraus, in denen es
keine festen Körper geben kann. Die Quantentheorie bedarf
also einer Reflexion ihrer geometrischen Voraussetzung.
Die dazu nötige Messtheorie erfordert eine Quantentheorie
vieler Objekte. Dies führt zur Forderung einer speziellen
Symmetriegruppe der Dynamik eines Gesamtobjekts, die alle
seine gleichartigen Teilobjekte gleichartig transformiert.
Wählt man als gleichartige Teilobjekte die einfachsten
möglichen, mit zweidimensionalem Zustandsraum ("Urobjekte"
= "Ure"), so ist die resultierende Gruppe die SU (2). Die
Zustandsvektoren jedes zusammengesetzten Objektes müssen
dann einen Darstellungsraum dieser Gruppe bilden. Sie
werden dazu zweckmässig als Funktionen auf die Gruppe ge-
schrieben. Die Gruppe, als Raum betrachtet (sie ist ihr
eigener grösster homogener Raum) fungiert dann als Ortsraum.
Auf diese Weise lässt sich die ausgezeichnete Beschreibungsweise der Physik durch einen dreidimensionalen reellen gekrümmten Ortsraum begründen. Raum- und Zeitmetrik basieren
in dieser Theorie auf der Wahrscheinlichkeitsmetrik des
Zustandsraums des Urs. Orte und Zeiten sind also keine
Observablen des Urs, sondern nur für grosse Anzahlen von
Uren statistisch messbar; sie sind klassische Grössen.

REFLEKTIERENDER RÜCKLAUF:

10'. Quantentheorie des Raums. Wenn diese Theorie durchführbar ist, so ist der Raum kein der Physik hierarchisch vorge-
ordnetes Datum, sondern eine genäherte Beschreibungsweise
zusammengesetzter Objekte. Der Fundamentalbegriff, der eine
Bedingung jeder möglichen Erfahrung formuliert, ist in dieser Theorie der Begriff der empirisch entscheidbaren Alternative. Er setzt ein Verständnis der Zeitmodi in der Gestalt
der faktischen Vergangenheit und der möglichen Zukunft voraus, aber schon keinen Begriff von Zeitpunkten. Die Theorie
liefert Metrik, Topologie und Dimensionzahl des Raumes.
Diese Theorie ist bisher nicht durchgeführt, aber ihre
Denkmöglichkeit wirft ein verändertes Licht auf die erkenntnistheoretischen Prämissen der bisherigen Theorien.

9'. Allgemeine Relativitätstheorie. Die Bedeutung der
Näherung getrennter Objekte, auf der die lokale Auszeichnung
der Poincarégruppe beruht, wurde durch die Annahme der Ure
aufgeklärt. Die Quantentheorie der Elementarteilchen ist
die bei Einstein fehlende Theorie der Materie. Die Theorie
der Ure liefert Topologie und Dimensionszahl des Weltkon-
tinuums. Nur die lokalen Variationen der Raumkrümmung, also
die Gravitationen im engeren Sinne, wären dann Ausdruck
desjenigen Teils der gesamten Wechselwirkung, der sich als
Raumstruktur darstellen lässt.

8'. Spezielle Relativitätstheorie. Als Theorie der Poincaré-
gruppe ist sie der ARth methodisch ebenso vorgeordnet wie
die Theorie freier Objekte der Theorie der Wechselwirkung.

7'. Semantische Konsistenz. Die skizzierte Theorie wäre
dem Ideal der semantischen Konsistenz näher als irgendeine
bisherige, damit auch dem Ideal einer einheitlichen Physik.
Soweit sie nur das Begriffsfeld voraussetzt, das zum Verständnis des Begriffs der empirisch entscheidbaren Alternative gehört, ist sie eine Art Erfüllung des Programms,
allgemeine Gesetze auf Bedingungen der Möglichkeit von
Erfahrung zu begründen. Sie würde damit zugleich zeigen,
inwiefern sowohl alle früheren Versuche, dies zu leisten,
als auch alle früheren Versuche, ohne dies auszukommen, an
bestimmten Stellen scheitern mussten. Sie würde die aus
empirischem Anlass gemachten, aber durch einen solchen
Anlass nie in Strenge gerechtfertigten Hypothesen allgemeiner Gesetze nachträglich aus den Bedingungen der Erfahrung

rechtfertigen. Selbst wenn sie durchführbar ist, stösst ihre
semantische Konsistenz aber an zwei Schranken. Erstens müsste
sie durch eine Theorie der Subjekte ergänzt werden, welche
die Subjekte nicht durch Messapparate ersetzt, deren Ables-
barkeit schlicht vorausgesetzt wird; d.h. eine in weiterem
Sinne semantisch konsistente Naturwissenschaft müsste die
Biologie und Anthropologie mit enthalten. Zweitens ist die
Annahme der empirischen Entscheidbarkeit irgendeiner Alternative stets nur eine Näherung. D.h. die objektivierende
Naturwissenschaft ist als solche nur ein Näherungsverfah-
ren. Es bleibt hier offen, wie weit wir jenseits dieses
Verfahrens zu denken vermögen, auch wie weit es selbst
Ausdruck einer geschichtlichen Entwicklungsphase des mensch-
lichen Bewußtseins ist.

6'. Operative Begründung der euklidischen Geometrie. Durch
die Quantentheorie kennen wir Recht und Grenzen der operativen Annahme über feste Körper. Es zeigt sich auch, dass
die Quantentheorie der einfachen Alternative eine tiefere
Begründung der fundamentalen Symmetriegruppen gibt. Die
festen Körper sind selbst schon spezielle Darstellungen
dieser Symmetrien.

5'. Konventionalismus. Es ist zu hoffen, dass das Problem
des geometrischen Konventionalismus durch die Auszeichnung spezieller Gruppen geklärt ist. Freilich ist in diesem Aufsatz der fundamentale Schritt, die Auszeichnung
der Wahrscheinlichkeitsmetrik in der Axiomatischen Quantentheorie,nicht behandelt.

4'. Kants Fragestellung. Der Kern der Kantschen Theorie,
die transzendentale Einheit der Apperzeption und die Selbst-
unterscheidung des Ich in transzendentales und empirisches
Subjekt würde erst in der Theorie der Subjekte (7'.) zur
Sprache kommen. Die These, Raum und messbare Zeit seien
das klassische, d.h. nach Bohr beobachtbare Schema der
Darstellung beliebiger zusammengesetzter Objekte, erinnert
an Kants Theorie, sie seien die Formen aller Anschauung.
Diese These vermeidet den unkantischen Nebenklang, die
Formen der Anschauung hätten etwas mit der speziellen
Beschaffenheit der Sinnlichkeit der Spezies Mensch zu tun;
sie hält nur den kantischen Bezug auf bewusste endliche
Subjekte überhaupt fest. Sie beansprucht auf dem für Kant
unerreichbaren Weg über die Quantentheorie die mathema-
tische Struktur des Raum-Zeit-Kontinuums (also, kantisch

gesagt, den Beitrag des Verstandes zur reinen Anschauung)
zu begründen.

3'. Hierarchismus. Er ist die Hoffnung, Recht und Grenzen
des Hierarchismus mit Begriffen aus dem Umkreis der semantischen Konsistenz zu erläutern. Das hierarchisch Vorge-
ordnete war in den historischen Beispielen meist ein Element
des Vorverständnisses, das zugleich als der speziellen
Theorie logisch übergeordnet verstanden wurde. So sagte
man, alle physischen Vorgänge seien im Raum, also gehe die
Theorie des Raums der Physik voran. Das ist nach der hier
vertretenen Auffassung auch in einer semantisch konsistenten Quantentheorie des Raumes in gewissem Sinne richtig.
Aber dass diese Theorie des Raumes eine Quantentheorie
beliebiger zusammengesetzter Objekte ist, konnte man
historisch in der Phase der klassischen Physik nicht wissen.
Das sachlich Übergeordnete erscheint in einer solchen
früheren Phase nicht in seinem wahren, notwendigen Zusammenhang mit dem ihm Untergeordneten, und eben darum erweckt
es den Anschein einer Notwendigkeit a priori, die doch
nicht ausgewiesen werden kann. Eine tiefergehende Analyse
solcher Zusammenhänge müsste zunächst eine Theorie der
Logik und ihres Zusammenhangs mit der Erfahrung (Logik
zeitlicher Aussagen) aufstellen.

2'. Empirismus. Dass man allgemeine Sätze nicht empirisch
begründen kann, ist seit Popper, eigentlich seit Hume, in
Wahrheit seit Platon bekannt. Andererseits ist es der
historische Hergang, dass sie ständig so begründet werden;
in der Darstellung der beiden Relativitätstheorien habe
ich versucht, gerade dies zu pointieren. Die Erfahrung
motiviert allgemeine Hypothesen, die, wenn sie wahr sind,
die empirischen Fakten nachträglich als notwendig erweisen.
Poppers Falsifikationsthese ist undeutlich, da Falsifikation selbst schon Theorie voraussetzt; sonst könnte man
ja z.B. die euklidische Geometrie durch ein Experiment
falsifizieren. Die von Kuhn beschriebene Paradigmenfolge
funktioniert, wie Kuhn weiss, nur, wenn die Natur sich
"gesetzmässig verhält". Der erfolgreiche Empirismus der
Physiker hat die simpel realistische Struktur, dass man
an wirkliche Gesetze glaubt, und diese zu erraten sucht.
Die hier entworfene Theorie tendiert dahin, das Staunen
über 1 000 spezielle oder 6 allgemeine Gesetze auf das
Staunen über eine Grundtatsache zu reduzieren, dass nämlich Erfahrung möglich ist. Das Idealschema ist: Wenn

Erfahrung möglich ist, d.h. wenn prüfbare Prognosen über
präzisierte Alternativen möglich sind, dann ist ihre Theorie die semantisch konsistente Quantentheorie einschliesslich der Quantentheorie des Raums. Dann brauchen wir uns
nicht zu wundern, dass die Konsequenzen dieser Theorie für
spezielle Fälle empirisch in eben diesen Fällen gefunden
wurden. Das philosophische Staunen träte mit dieser Fest-
stellung nur in eine neue Phase.

1'. Axiomatik. Die Auswahl der empirisch relevanten Axiome
dürfte zur Genüge besprochen sein. Der Grund der Möglichkeit einer Axiomatik, einschliesslich des geometrischen
Vorverständnisses der Metamathematik, ist Gegenstand einer
Theorie der Logik und Mathematik.

ANMERKUNGEN

1) Imro Tóth 1967, Das Parallelenproblem im Corpus Aristo-
   telicum. Archive for the History of Exact Science, 3,
   S. 249-421.

   Hugo Dingler 1938, Die Methode der Physik, München;
   Aufbau der exakten Fundamentalwissenschaft, hrsg. v.
   Paul Lorenzen, München 1964.

2) Nach Abschluss dieser Arbeit erhielt ich das Buch von
   Jörg Winkler, "Relativität und Eindeutigkeit, Hugo
   Dinglers Beitrag zur Begründungsproblematik", Monogra-
   phien zur Philosophischen Forschung, Band 98, Meisen-
   heim am Glan 1973.

   Thomas S. Kuhn 1962, "The Structure of Scientific
   Revolution", Chicago.

   W. Heisenberg in Dialectica 1948, S. 331-336 und in
   "Einheit und Vielheit", Göttingen 1972, S. 140-144.

3) Zur Axiomatik dieser Geometrie s. H. Ehlers in
   "The Physicist's Conception of Nature", ed. J. Mehra,
   Reidel, Dordrecht, 1973, S. 70-91.

4) P.A.M. Dirac, Phys. Zeitschr. der Sowjetunion 3, Heft 1,
   1933. Dazu R.P. Feynman, Rev. of Mod. Physics 20, 267,
   1948. Beide abgedruckt in "Quantum Electrodynamics",
   ed. J. Schwinger, Dover 1958.

   Joseph M. Jauch 1968, "Foundations of Quantum Mechanics",
   Addison-Wesley.

5) M. Drieschner, Dissertation Hamburg 1968. Dazu C.F. v.
   Weizsäcker, "Die Einheit der Natur", München 1971, II,
   5 und l.c.3) S. 635-667

6) C.F. v. Weizsäcker, l.c.3), S. 55-59.

Heinz–Jürgen Schmidt

KINEMATICS AS A THEORY OF COINCIDENCES

0. INTRODUCTION

In this article, "kinematics" will be considered as a pre-theory of mechanics dealing with macroscopic bodies in the laboratory. An axiomatic account of such a pre-theory should meet the following requirements:

(i)   The basic concept should be closely related to the intended range of application – the movement of bodies; no reference to light rays or atomic clocks is desirable and necessary.

(ii)   The notion of an inertial system should be defined, if the axiomatics depends on it.

(iii) Though relativistic space–time can be characterized by using light rays only (cf. Robb (1936)) it would be instructive to give a unified treatment of relativistic and non–relativistic kinematics.

(iv)   The approach should be synthetic. The use of space–time coordinates is a practical mean to represent kinematical structure but tends to veil the physical core.

(v)   The role of idealization and the underlying uniform structure should be clarified.

Though the present paper is guided by these principles, the reader will soon recognize that they are only partially satisfied. Furthermore, we are not dealing with general kinematics but only with the theoretical representation of uniform motion. Let me explain the basic ideas, which surely may be traced back to other treatments on the same subjects (especially Reichenbach (1928), Robb (1936), Freudenthal (1974), Havas (1964), Lévy-Leblond (1969), Brennich (1969), Süßmann (1969), Ehlers et al. (1972), Ehlers (1973), Schutz (1973), Ludwig (1974), Mayr (1979).
Usually the mathematical description of kinematics is

D. Mayr and G. Süssmann (eds.), Space, Time, and Mechanics, 87–104.
Copyright © 1983 by D. Reidel Publishing Company.

based on a set E of _events_, which is an affine 4-dimension-
al space. The further structure can be specified by means
of a group G, affinely acting on E, which is either the
Galilean group or the Poincaré group. But only if G is interpreted in the active sense, i.e. as a group of reproductions of processes (cf. (Mayr (1979)), it is suited as a
basic datum of axiomatics. If G is interpreted in the
passive sense, as a group of coordinate transformations, or,
equivalently, of automorphisms of a certain structure, it
is preferable to make this structure explicite. Following
Havas (1964), this can be done in the following way: Let V
be the vector space associated with E, V* its dual space
and $g : V \to V^*$, $h : V^* \to V$ be the two symmetric tensors of
rank 2, which are invariant under G. They are unique up to
a factor; let the corresponding 1-dimensional subspaces be
[g] and [h]. Then we may _define_ the kinematical structures,
which are the subject of our axiomatization, as follows:

D1: A kinematical structure is a triple (E,[g],[h]), where
E is a 4-dimensional real affine space, [g] resp. [h]
are 1-dimensional subspaces of co- resp. contra-variant
second rank tensors of one of the following types: With
respect to an appropriate coordinate system,

(i) $g = h^{-1}$ has components $(g_{\mu\nu}) = \begin{pmatrix} -1 & & & \\ & 1 & & \\ & & 1 & \\ & & & 1 \end{pmatrix}$
(_hyperbolic_ kinematics), or

(ii) $(g_{\mu\nu}) = \begin{pmatrix} 1 & & & \\ & 0 & & \\ & & 0 & \\ & & & 0 \end{pmatrix}$, $(h^{\mu\nu}) = \begin{pmatrix} 0 & & & \\ & 1 & & \\ & & 1 & \\ & & & 1 \end{pmatrix}$
(_parabolic_ kinematics), or

(iii) $g = h^{-1}$ has components $(g_{\mu\nu}) = \begin{pmatrix} 1 & & & \\ & 1 & & \\ & & 1 & \\ & & & 1 \end{pmatrix}$,
(_elliptic_ kinematics).

The latter (unphysical) possibility is considered only
by reason of symmetry. The corresponding automorphism group
will be

D2: $\tilde{G}$ = A u t (E,[g],[h]), containing translations, proper
spatial rotations, spatial reflections, time reflections
and
(i)   Lorentz boosts, 4-dimensional dilations, or
(ii) Galilean boosts, spatial dilations, time dilations, or

(iii) 4-dimensional rotations and dilations.

An axiomatization of kinematics has to explore the physical meaning of these structures. But this is possible only in an indirect way. To each inertial system i_1 there belongs an affine equivalence relation on E, whose fibers are 1-dimensional ("world-lines") and enclose all events happening at the same place ("point") with respect to i_1. We may write this in the form

$$\pi : E \to P, \tag{O.1}$$

where P is the set of points. P can be endowed with the structure of a Euclidean space by means of the tensors g and h: Using $g : V \to V^*$ we can assign to each inertial system i_1 a hyperplane H_1 in V, which is affinely isomorphic to P. The restriction of h to this hyperplane yields the Euclidean structure we were looking for. Therefore a more direct way of representing the kinematical structure would be to consider a set E, a Euclidean space P and a class P of projections $\pi : E \to P$ satisfying certain axioms.

However, we will use a slightly different approach due to the view that "events" and "points" are both idealizations of the more fundamental concept of a physical "process". In applications of kinematics we are concerned with movements of solid, approximately rigid bodies. These movements are examples of physical processes, which form a partially ordered set (Π, ζ) with respect to the inclusion ζ of processes. They could be visualized by world-tubes in a space-time diagram. If we study the possible arrangements of rigid bodies in relative rest, we obtain a geometry in one reference system (or frame), characterized by a group of transport mappings. A closer analysis shows that the process of abstraction from concrete bodies to geometrical entities depends on the absence of material-depending deformation. This enables us to give an intrinsic definition of inertial systems. For details see Schmidt (1979).

How are the geometries of different inertial systems related? A physically meaningful relation must be based on some interaction between bodies in rest with respect to different inertial system. Consider for instance the collosion between two billiard balls, the trace in snow leaved behind by a uniformly gliding sledge or the wheels of a

train keeping close contact with straight rails. These are
examples of what we shall call coincidences. If two bodies
come close together during their movement they will be said
to <u>coincide</u>. Similarly, 3 bodies "coincide" if they come
close together simultaneously. There are different ways to
express this phenomenon in the language of processes, for
instance by the assertion that 3 processes (movements of
bodies) have a common ζ-meet.

Let us pass to "idealized coincidence". If J is the
set of inertial systems, to each $i \in J$ we assign a geometrical space P_i, such that the position of bodies in i-rest
is represented by "regions", specific subsets of P_i. Certain limits of smaller and smaller regions are called
<u>points</u> of P_i and we may extend the coincidence relation to
points, say

$$C(p_i, p_j, p_k), \text{ where } p_i \in P_i, p_j \in P_j, p_k \in P_k, i, j, k \in J \quad (0.2)$$

if p_i, p_j, p_k coincide. This is the only concept, additional to the geometrical ones, which we need to characterize
physical kinematics. For instance the law of inertia may be
reformulated as the postulate that the sets of pairs
(p_i, p_j) such that $C(p_i, p_j, p_k)$ with a fixed p_k form a
"straight line" in $P_i \times P_k$ (for our definition of "straight
line" cf. section 4). Of course, here we neglect gravitational effects. Allowing for these effects would require
radical changes in the underlying geometrical structure.

We may thus anticipate one result of this article: it
is not necessary to introduce the concept of a 'clock'
additionally to 'meter-stick' and 'inertial system'. Clocks
and synchronization procedures may be constructed from uniformly moving bodies. A similar idea has been used by P.
Janich (1969) in his chronometry. However, in contrast to
that approach, the space-time structure characterized in
this article is considered to be open for alterations and
refinements by more comprehensive theories.

1. GEOMETRY

I will stick to my (1979) axiomatic characterization of
physical geometry based on the solution of the Helmholtz-
Lie space problem due to Tits (1955) and Freudenthal (1956).

Consider a set of points P, equipped with a topology τ and a subgroup T of Aut(P,τ). Points and (a base of) τ-open subsets are physically interpreted as idealizations of small rigid bodies, affixed to certain references bodies "in rest", and mappings contained in T as transports of these bodies (see also Ludwig (1974). It is possible to give a physical meaning with respect to this interpretation to the following postulates:

AXIOM G': (P,τ) is a locally compact Hausdorff space; T is connected, complete (with respect to the two-sided uniformity of pointwise convergence) and operates transitively, rigidly and mobilly.

"Rigid operation" means, that for any A, B, A closed, B compact, there exists an open U such that tU, $t \in T$, never meets both A and B. "Mobility" is understood in the sense that for two points x, y $\in$ P the set
$P \smallsetminus \{z \in P \mid \exists t \in T : z = ty$ and $tx = x\}$ is not connected.

Under these assumptions P is shown to be an n-dimensional Riemannian space of one of the following types:

Either a parabolic (= affine Hilbert) or hyperbolic or elliptic or spherical space over the (skew-) field of the real or complex or quaternion or Cayley (n = 2) numbers. The group T is essentially the automorphism group of P.

The spaces of real dimension 3 enjoy the properties of full mobility and constant curvature. However, we shall first stick to the general situation of axiom G'. The (real) dimension of P is only supposed to be greater that 1.

For later purposes it will be convenient to mention that in the real parabolic (= Euclidean) case, Aut(P,τ,T) is generated by translations, proper rotations, dilations, and reflections. Usually this fact is referred to as homogenuity and isotropy of space and, further, as the lack of a natural unit of length and of a natural orientation of space.

The reconstruction of primitive geometry as a theory of operations with rigid bodies reveals a tacit assumption

of great conceptual relevance: the transport of spatial
regions must not depend on the material of the bodies
which represent the region. Material-dependent deformations
of the bodies would make impossible a unique definition of
a transport mapping. But this independence cannot be assumed unless the system defined by the used reference bodies
is, approximately, an <u>inertial system</u>. (Of course, the required degree of inertiality depends on the rigidity of the
used bodies.) Without expanding this idea we may state,
that

1. there are intrinsic criteria for free motion of reference bodies, thus defining inertial systems, which are

2. constitutive for primitive geometry of rigid bodies, and

3. to each inertial system we must assign its own geometry.

Taking into account this variety of inertial systems,
a physical geometry is now characterized by a 5-typel

D3: $(J,P;\pi,\tau,T)$, where

$$P = \bigcup_{i \in J} P_i$$

$\pi : J \to \text{Pot } P$, such that $\pi(i) = P_i$,

$\tau = (\tau_i)_{i \in J}$, $T = (T_i)_{i \in J}$.

Here J denotes the class of inertial systems and (P_i,τ_i,T_i)
is the geometry belonging to the inertial system i. Thus
we postulate:

<u>AXIOM G</u>: Each geometry (P_i,τ_i,T_i), $i \in J$, satisfies axiom
G'. Its dimension is ≥ 2.

<u>AXIOM J</u>: There are at least 2 different inertial systems.
Any two geometries P_i, P_j, $i,j \in J$, are isomorphic.

These two axioms together with D3 obviously define a
<u>species of structure</u> Σ in the sense of Bourbaki (1968).

2. COINCIDENCES

We will axiomatize the relation, that the processes $P_1 \ldots P_n$
have a non-void intersection, for the limiting case of
$P_1 \ldots P_n$ being geometrical points of possibly different
inertial systems. This relation will be called <u>coincidence</u>

relation. It will be enough to consider 3-point coincidence, but 2-point coincidence fails to be transitive and hence doesn't suffice to regain the general case.

Thus let

D4: $C \subseteq P \times P \times P$

and write $C(p_1,p_2,p_3)$ for $(p_1,p_2,p_3) \in C$, further:

$C(p_1,p_2) \underset{\text{def}}{\leftrightarrow} \exists p_3 \in P : C(p_1,p_2,p_3)$,

and, if $n > 3$, $C(p_1 \ldots p_n) \underset{\text{def}}{\leftrightarrow} \forall i,j,k \in \{1 \ldots n\}$:

$C(p_i,p_j,p_k)$.

Sometimes it will be convenient to write $C(1,2,3)$ for $C(p_1,p_2,p_3)$ etc.

The following postulates are rather simple, but note that (iii) is connected with time being one-dimensional.

AXIOM $\overline{C}$:

(i) $C(p_1,p_1,p_1)$.

(ii) $C(p_1,p_2,p_3)$ is a totally symmetric relation.

(iii) Let $i \in J$ and $C(p_1,p_2,p_3)$. Then there exists a $p_i \in P_i$ such that $C(p_1,p_2,p_3,p_i)$. p_i is unique, if not $p_1 = p_2 = p_3$.

(iv)   If $i \neq j \in J$ and $p_j \in P_j$, then $|\{p_i \in P_i | C(p_i,p_j)\}| \geq 2$.

Lemma 1: $C(1,2) \Rightarrow C(1,2,1)$.

Proof: $\exists 3 \in P_3$ such that $C(1,2,3)$ by def. of $C(1,2)$. By Axiom C (iii) $\exists 1' \in P_1$ such that $C(1,2,3,1')$. $1 \neq 2$ implies $1 = 1'$ by uniqueness, hence $C(1,2,1)$. If $1 = 2$, Axiom C (i) applies. $\square$

T1: Let $n \geq 3$, $i \in J$ and $C(p_1 \ldots p_n)$. Then there exists $p_i \in P_i$ such that $C(p_1 \ldots p_n, p_i)$. p_i is unique, if not $p_1 = \ldots = p_n$.

Proof: If $n = 3$, this is just Axiom C (iii). Let us assume T1 as induction hypothesis for $n = 3 \ldots k$ and prove it for $n = k + 1$. Choose p_i such that $C(p_1 \ldots p_k, p_i)$. We may assume that all p_j, $j = 1 \ldots k + 1$, are different, otherwise $n \leq k$. Hence p_i is unique. It remains to show that $C(p_i, p_j, p_1)$ for $j, 1 \in \{1 \ldots k+1\}$. The case $j, 1 < k + 1$ is covered by in-

duction hypothesis. Let, without loss of generality,
$1 < j < l = k + 1$ and choose $q_i \in P_i$ such that
$C(p_2 \ldots p_1, q_i)$. By uniqueness (and $k \geq 3$), $p_i = q_i$, hence
$C(p_j, p_1, p_i)$.   □

T2: If $3 \neq 4$, $C(1,2,3,4)$ and $C(3,4,5,6)$, then $C(1,2,5,6)$.

Proof: By T1 $\exists$ 5' $\in P_5$, 6' $\in P_6$ such that $C(1,2,3,4,5)$ and
$C(1,2,3,4,5',6')$. Since $3 \neq 4$, Axiom C (iii) implies $5 = 5'$
and $6 = 6'$.   □

D5: (i) $C_{\neq} \underset{\text{def}}{=} \{(p_i,p_j) \in P \times P \mid i \neq j \text{ and } C(p_i,p_j)\}$.

   (ii) $(1,2) \; \kappa \; (3,4) \underset{\text{def}}{\Leftrightarrow} C(1,2,3,4)$.

T3: κ is an equivalence relation on $C_{\neq}$.

Proof: (i) $C(1,2)$ implies $C(1,2,1)$ and $C(1,2,2)$ by lemma 1.
Hence $C(1,2,1,2)$ and $(1,2) \; \kappa \; (1,2)$. Thus κ is reflexive.

(ii) κ is symmetric by Ax. C (ii).
(iii) κ is transitive be T2.   □

D6: (i)   The elements of $E \underset{\text{def}}{=} C_{\neq}/\kappa$ will be called <u>events</u>.
         Sometimes they will be denoted by
         $e = [p_1,p_2] \in E$ if $(p_1,p_2) \in C_{\neq}$.

   (ii)   Let us define maps
         $f_i : E \to p_i$, $i \in J$, by $P_i = f_i[p_1,p_2] \Leftrightarrow C(p_1,p_2,p_i)$.
         P_i is unique by virtue of Ax. C (iii) and lemma
         1.

   (iii) $f_i^{-1}(p_i)$ will be called the <u>world line</u> of $p_i \in P_i$.

   (iv)   Let $i \neq j \in J$ and $C_{ij} \underset{\text{def}}{=} \{(p_i,p_j) \in P_i \times P_j \mid C(p_i,p_j)\}$.
         $f_{ij} : E \to C_{ij}$ will be defined by
         $f_{ij}(e) = (f_i(e), f_j(e))$. $f_{ij}(e) \in C_{ij}$ follows from
         $C(p_1,p_2,p_i)$, $C(p_1,p_2,p_j)$ and the uniqueness of
         $q_j \in P_j$ satisfying $C(p_1,p_2,p_i,q_j)$.

T4: (i) $f_{ij} : E \to C_{ij}$ is bijective.
   (ii) $C(p_1,p_2,p_3) \Leftrightarrow f_1^{-1}(p_1) \cap f_2^{-1}(p_2) \cap f_3^{-1}(p_3) \neq \emptyset$.

Proof: (i) The map f_{ij} picks up just one element (p_i,p_j)
from each equivalence class $e \in C_{\neq} / \kappa$.

(ii) Let $C(p_1,p_2,p_3)$, then we have $p_i = f_i[p_1,p_2]$ for
$i = 1,2,3$ and $[p_1,p_2] \in \bigcap_{i=1,2,3} f_i^{-1}(p_i)$. Conversely, let
$[p_4,p_5] \in \bigcap_{i=1,2,3} f_i^{-1}(p_i)$. Then it follows, that
$C(4,5,1)$, $C(4,5,1,2')$ and $C(4,5,2)$, hence $2 = 2'$. Equally
$C(4,5,1,2,3')$ and $C(4,5,3)$ imply $3 = 3'$, hence $C(1,2,3)$. □

.3. KINEMATICAL AUTOMORPHISMS AND THE PRINCIPLE OF RELATIVITY

The <u>species of kinematical structure</u> $\sum_k$ will be obtained by
expansion of $\sum$; we add the term $C \subset P \times P \times P$ to the
structural term of $\sum$ and obtain:

D7: (i)   Basis sets: J,P,

    (ii)   structural term (π,τ,T,C) with obvious typificat-
           ion,

    (iii) $k = (J,P;\pi,\tau,T,C)$ is a structure of species $\sum_k$,
           iff the axioms G, J, C, R, I, S (see below) are
           satisfied.

    In accordance with the general definition of automor-
phisms (Bourbaki (1968) IV § 1.5) <u>kinematical automor-</u>
<u>phisms</u> will take the following form:

T5: $\psi \in A u t(k) \leftrightarrow \psi = (\sigma,(\psi_i)_{i\in J})$, where $\sigma : J \rightarrow J$ is a
     permutation of inertial systems and $(\psi_i:P_i \rightarrow P_{\sigma i})_{i\in J}$ is a
     family of geometrical isomorphisms. Moreover, for all
     i,k, $k \in J$, $C(p_i,p_j,p_k) \leftrightarrow C(\psi_i p_i,\psi_j p_j,\psi_k p_k)$ is valid.

    Now we may formulate Einstein's principle of relati-
vity - including conservation of geometrical symmetry - in
the following way:

<u>AXIOM R:</u> If $i,j \in J$ and $\psi : P_i \rightarrow P_j$ is a geometrical iso-
morphism, then there exists a $\tilde{\psi} = (\tilde{\sigma},(\psi_i)_{i\in J}) \in A u t(k)$
such that $j = \sigma i$ and $\psi = \psi_i$.

    Note that Axiom R comprises two postulates: First,
$A u t(k)$ operates transitively on J. This implies that any
two inertial systems have equal status, which is the proper
principle of relativity. Secondly, each geometrical <u>auto-</u>

morphism $\psi : P_i \to P_i$ can be extended. Hence kinematics does not enrich the geometrical structure, which is just the kind of symmetry argument often used in the foundations of special relativity.

Kinematical automorphisms operate on E in a natural fashion. Let $\psi = (\sigma, (\psi_i)_{i \in J}) \in A u t(k)$, then we define:

D8: $\psi^E : E \to E$ by $\psi^E[p_1, p_2] \underset{\text{def}}{=} [\psi_1 p_1, \psi_2 p_2]$.

By reason of T5 we have $C(p_1, p_2) \leftrightarrow C(\psi_1 p_1, \psi_2 p_2)$, hence $(\psi_1 p_1, \psi_2 p_2) \in C_+$ and $[\psi_1 p_1, \psi_2 p_2] \in E$. The definition D8 does not depend on the choice of the element from the equivalence class $[p_1, p_2]$, because $C(p_1, p_2, p_3, p_4) \leftrightarrow C(\psi_1 p_1, \psi_2 p_2, \psi_3 p_3, \psi_4 p_4)$ holds. The following theorem is immediate:

T6: ψ^E is a bijection of E and $\psi \to \psi^E$ is an injective group homomorphism $A u t(k) \to B i j(E)$. ψ^E generates a bijection of the set of world lines of points $p \in P_i$ onto the set of world lines of points $q \in P_{\sigma i}$.

4. LAWS OF INERTIA AND SUPERPOSITION

A point in the system $k \in J$ will leave "traces" in other systems $i, j \in J$. The law of inertia tells us that these traces are straight lines, which could be translated by "geodesics" in the general geometrical setting. However, we will consider a larger class of "nice" curves in order to reformulate this law, namely orbits of 1-parameter subgroups of transport mappings. We need some definitions to express the corresponding axiom.

D9: Let $i \neq j$, $k \in J$, $p_k \in P_k$.

(i) $s_{ij}^k(p_k) \underset{\text{def}}{=} \{(p_i, p_j) \in C_{ij} | C(p_i, p_j, p_k)\}$,

(ii) $s_{ij}^k \underset{\text{def}}{=} \{s_{ij}^k(p_k) | p_k \in P_k\}$,

(iii) $s_{ij} \underset{\text{def}}{=} \{s_{ij}^k | k \in J\}$.

Similarly,

(iv) $s_i^k(p_k) \underset{\text{def}}{=} \{p_i \in P_i | C(p_i, p_k)\}$,

(v) $S_{i\ \text{def}}^{k} = \{S_i^k(p_k)\,|\,p_k \in P_k\}.$

(vi) $S_{i\ \text{def}} = \{S_i^k\,|\,k \in J\}$

D10: (i) T_i will denote the Lie algebra of the group T_i (which is a Lie group, cf. Schmidt (1979),(425)).

(ii)   If $t_i \in T_i$, $t_j \in T_j$, then
$$<t_i,t_j>_{\text{def}} = \{\{((\exp \lambda\, t_i)p,(\exp \lambda\, t_j)q)\,|\,\lambda \in \mathbb{R}\}\,|$$
$$p \in P_i, q \in P_j\}.$$
$<t_i,t_j>$ is the family of orbits of a 1-parameter subgroup of $T_i \times T_j$.
Similarly,

(iii) $<t_i>_{\text{def}} = \{(\exp[t_i])p\,|\,p \in P_i\}$, where

(iv) $[t_i]_{\text{def}} = \{\lambda\, t_i\,|\,\lambda \in \mathbb{R}\}.$
$[T_i]_{\text{def}} = \{[t_i]\,|\,t_i \in T_i\}$ denotes the projective space generated by T_i.

(v) $<T_i>_{\text{def}} = \{<t_i>\,|\,t_i \in T_i\},$
$$<T_i,T_j>_{\text{def}} = \{<t_i,t_j>\,|\,t_i \in T_i, t_j \in T_j\}.$$

(vi) $I_{i\ \text{def}} = \{s \in T_i\,|\,\exists k \in J: <s> = S_i^k\}.$

Let us state without proof the following proposition:

T7: $<s> = <t> \Leftrightarrow [s] = [t].$

The nontrivial direction may be proved by using the differential equation for killing vector fields (= infinitesimal isometries) (cf. Petrov (1969), 10.11, or Kobayashi/Nomizu (1963), VI-3)

AXIOM 1: If $i \neq j \in J$, then $S_{ij} \subset <T_i,T_j>$, but $S_i \subsetneqq <T_i>$.

Hence inertial orbits are generated by 1-parameter subgroups of transport mappings, but not all 1-parameter-group generate such orbits. Next we postulate some kind of superposition principle for "inertial vectorfields":

$\underline{\text{AXIOM } S:}$ If $\langle t_i^j \rangle = S_i^j$ and $\langle t_i^k \rangle = S_i^k$, then there exists some $1 \in J$ such that $\langle t_i^j + t_i^k \rangle = S_i^1$.

Lemma 2: I_i (cf. D10 (vi)) is a proper ideal of T_i.

Proof: By axiom I and S, I_i is a linear subspace of T_i, non-zero by axiom C (iv). By virtue of axiom R, I_i is invariant under inner automorphism of T_i. □

T8: (P_i, τ_i, T_i) is of parabolic type. S_{ij}^k is a maximal family of parallels in $C_{ij} \subset P_i \times P_j$.

Proof: In all other cases the Lie algebra T_i is (real) simple (cf. Freudenthal (1956), Satz 2). Hence I_i consists of translations and its orbits are parallel. □

5. AFFINE STRUCTURE OF E

For $i \neq j \in J$, S_i^j is a family of parallels in P_i, likewise S_j^i in P_j. Let us define a mapping

D11: $\Gamma_i^j : S_i^j \to S_j^i$ by $\Gamma_i^j S_i^j(p_j) = S_j^i(p_i)$ iff $p_i \in S_i^j(p_j)$.

Γ_i^j is well defined, for if $q_i \in S_i^j(p_j)$ we conclude $C(q_i, p_j)$ and $p_j \in S_j^i(q_i)$. But $S_j^i(p_i)$ and $S_j^i(q_i)$ are parallel and have the point p_j in common, hence they coincide.

Note, that S_i^j may be viewed as an $(n-1)$-dimensional affine quotient space of P_i. Let $\sigma_i^j : P_i \to S_i^j$ be the corresponding quotient map.

T9: $\Gamma_1^2 : S_1^2 \to S_2^1$ is an affine bijection.

Proof: Consider in P_1 the line $S_1^3(p_3)$ which may be taken non-parallel to the lines of S_{1}^2. Such a line exists by axiom R. Analogously consider $S_2^3(p_3)$ in P_2. We shall show: $\Gamma_1^2 \sigma_1^2 S_1^3(p_3) = \sigma_2^1 S_2^3(p_3)$. Let $p_1 \in S_1^3(p_3)$. There exists a unique $p_2 \in P_2$ such that $C(p_1, p_2, p_3)$, hence $\sigma_1^2 p_1 = S_1^2(p_2)$. Then we have $\Gamma_1^2 \sigma_1^2 p_1 = S_2^1(p_1) \in \sigma_2^1 S_2^3(p_3)$ since $p_2 \in S_2^3(p_3)$. The $\supset$-inclusion then follow from $(\Gamma_1^2)^{-1} = \Gamma_2^1$. Each line in the affine space S_1^2 is of the form $\sigma_1^2 S_1^3(p_3)$, because the stability subgroup $J_1 \subset T_1$ operates transitively on inertial lines in P_1. Thus Γ_1^2 maps lines onto lines and is therefore affine.   □

T10: C_{12} (cf. D6 (iv)) is an (n+1)-dimensional affine subspace of $P_1 \times P_2$.

Proof: Let $C(p_1,p_2)$, $C(q_1,q_2)$, $\lambda \in \mathbb{R}$ and
$r_i = \lambda p_i + (1-\lambda) q_i$ for i = 1,2. Then we must show:
$C(r_1,r_2)$. We have $C(p_1,p_2) \Leftrightarrow \Gamma_1^2 \sigma_1^2 p_1 = \sigma_2^1 p_2$ and
$C(q_1,q_2) \Leftrightarrow \Gamma_1^2 \sigma_1^2 q_1 = \sigma_2^1 q_2$. Since Γ_i^j and σ_i^j are affine
maps, we conclude $\Gamma_1^2 \sigma_1^2 r_1 = \sigma_2^1 r_2 \Leftrightarrow C(r_1,r_2)$.
The set $S_2^1(p_1)$ is a line in P_2. Therefore the pair
$(p_1,p_2) \in C_{12}$ is determined by p_1 and one further affine coordinate in P_2 increasing on $S_2^1(p_1)$. This implies that the
dimension of C_{12} is n + 1.   □

The affine structure on C_{12} may be transferred to E by
means of the bijection $f_{12} : E \to C_{12}$. In order to insure
that this structure doesn't depend on the choice of the
inertial systems we will show that $f_{34} \circ f_{12}^{-1}$ is affine.

We know from axiom I and T8 that $S_{12}^3(p_3)$ is a line in
$P_1 \times P_2$, hence in C_{12}. If S_{12}^3 is the corresponding family of
parallels, $\sigma_{12}^3 : C_{12} \to S_{12}^3$ the affine quotient map,
$\Gamma_{12}^3 : S_{12}^3 \to P_3$ the natural bijection, we have $\Gamma_{12}^3 \sigma_{12}^3 =$
$f_3 f_{12}^{-1} = g_3$. Similarly as in T9, Γ_{12}^3 is shown to be affine.
Therefore g_3 and $g_3 \times g_4 \overset{\text{def}}{=} f_{34} f_{12}^{-1}$ are affine maps too and
we have the following theorem.

T11: E can be endowed with the structure of an (n+1)-di-
mensional affine space such that the bijections
$f_{ij} : E \to C_{ij}$ (i,j $\in J$) become affine. World lines of
different points p, q $\in P_i$ are different parallels in
E. Each event e $\in$ E is located on such a parallel for
given i $\in J$. Two inertial systems with the same family
of world lines are equal.

The latter is true, because otherwise different
systems would produce trivial inertial lines $S_i^j(p_j)$, namely
points, which is excluded by axiom C (iv).

T12: The bijection $\psi^E : E \to E$, $\psi \in A u t(k)$, is affine.

Proof: Let $\psi = (\sigma, (\psi_i) i \in J)$, then we have
$$\psi^E = f^{-1}_{\sigma i, \sigma j} \circ (\psi_i \times \psi_j | C_{ij}) \circ f_{i,j}. \qquad \square$$

6. REPRESENTATION OF KINEMATICS

Let us fix an inertial system $i_1 \in J$ with corresponding geometry (P_1, τ_1, T_1). The subscript (1) will be occasionally omitted.

D12: $C_{(1)}$ denotes the subgroup of $A u t(k)$ of transformations satisfying $\psi_1 = Id_{P_1}$. $C^E \underset{def}{=} \{\psi^E | \psi \in C\}$. The elements of C resp. C^E, will be called <u>special Carroll transformations</u> (cf. Lévi-Leblond (1965)).

T12: C^E contains the subgroup $V_{(1)}$ of ("time-like") translations into the direction of world lines $f_1^{-1}(p_1)$, $p_1 \in P_1$.

Proof: V generates kinematical automorphisms such that $\sigma = Id_J$ and $\psi_i : P_i \to P_i$ is a translation. $\psi_1 = Id_{P_1}$ holds, since the world lines of points $p_1 \in P_1$ are invariant under V. $\square$

On account of this proposition we may restrict ourselves to Carroll transformations leaving fixed an event $o \in E$. Let us call this subgroup C^E_o.

Now consider an affine coordinate system in E with origin o, whose 1-axis is parallel to the world lines $f_1^{-1}(p_1)$, $p_1 \in P_1$. Such a system will be called P_1-system. Hence the coordinates $x_2 \ldots x_{n+1}$ may be considered as spatial coordinates in P_1. Let $c \in C^E_o$ and $e_i \in E$, $i = 1 \ldots n+1$, events with coordinates $x_j(e_i) = \delta_{ij}$. The events ce_i must lie on the same world lines as e_i, hence

$$x_j(ce_i) = \delta_{ij} + a_i \delta_{j1} \text{ for } i = 2 \ldots n+1,$$
$$x_j(ce_1) = a_1 \delta_{j1}, \ a_1 \neq o. \tag{1}$$

Thus c has a matrix with respect to this coordinate system of the form

$$\underline{c} = \begin{pmatrix} a_1 & a_2 \cdots a_{n+1} \\ & 1 \\ & & \diagdown \\ & & & 1 \end{pmatrix} \quad , \; a_1 \neq o, \tag{2}$$

which will be written in short-hand notation as

$$\underline{c} = (a_1, \vec{a}). \tag{3}$$

Further let

$$\underline{c}_o^E \underset{\text{def}}{=} \{\underline{c} \mid c \epsilon c_o^E\} \text{ with respect to a fixed } P_1\text{-system.} \tag{4}$$

Lemma 3: The subgroup of c_o^E consisting of transformations such that $a_1 = \pm 1$ in (2) has at most two elements.

Proof: Because $(-1,\vec{a})\;(-1,\vec{b}) = (1,\vec{b}-\vec{a})$, it suffices to show that $a_1 = 1$ holds only for the identity in c_o^E. In an appropriate P_1-system the transformation in question has a matrix (2) with $a_1 = 1$, $a_2 = a$, $a_3 = \ldots a_{n+1} = o$. If $a = o$ we are done. Thus assume $a \neq o$ and choose an inertial system $i_2 \in J$ whose world lines have the direction $(x,o,1,o,\ldots)$. For its existence we employ the argument of isotropy and $\dim P_1 \geq 2$. We may use as affine coordinates of P_2 the coordinates $x_2 \ldots x_{n+1}$ of the intersection of its world lines with the $x_1 = o$ hyperplane in E. The above matrix leaves the vector $(x,o,1,o,\ldots)$ invariant, hence $\psi_2 : P_2 \rightarrow P_2$ and its matrix can be calculated. It turns out to be diag $(\frac{x}{x+1},1,1\ldots)$. This is not a geometrical isomorphism and hence we have obtained a contradiction. $\square$

It will be enough to sketch the remaining arguments leading to the desired representation theorem.
Lemma 3 implies that $(a_1,\vec{a}) \in c_o^E$, $a_1 > o$, lies in the range of a 1-parameter subgroup $\exp(\lambda b_1, \lambda b)$ and is hence represented by matrices of the form diag$(a_1,1,\ldots 1)$ with respect to a suitable P_1-system. This easily extends to the whole c_o^E. If $c_o^E \neq \{1\}$, this P_1-system yields a unique hyperplane

$$H_1 \underset{\text{def}}{=} \{h \epsilon E \mid x_1(h) = o\}. \tag{5}$$

Now consider the rotation group J_o in P_1 and the group G_o of its possible extensions to E. We have $J_o \overset{\sim}{=} G_o/C_o$, thus $C_o = \{1\}$ implies complete reducibility of the representation $J_o \rightarrow G_o$, J_o being compact, which gives us the hyperplane H_1 in this case too. Let again $C_o \neq \{1\}$ and note that G_o operates on the set of orbits of its invariant sub-group C_o. H_1 consists of all 1-point orbits and is hence invariant under G_o. So in either case J_o is represented by matrices $R = \frac{1}{R}$, $R \in SO (3)$. H_1 becomes a Euclidean space and if $i_2 \in J$, the two Euclidean structures of $H_1 \cap H_2$ coincide because each rotation on $H_1 \cap H_2$ has two extens-ions to geometrical automorphisms of P_1 and P_2.

Let the world-lines of another inertial system $i_2 \in J$ be generated by the vector $(1,\vec{v})$, resp. $(0,\vec{v})$, then $(1,R\,\vec{v})$, resp. $(0,R\,\vec{v})$, again generates world-lines of an inertial system $i_3 \in J$. Repetition of this argument will show that J, considered as a subset of the projective space E_p, is open and connected. With respect to the fixed P_1-system all "velocities" $\vec{v}$ with the norm

$$|\vec{v}| < c, c > o, \text{ or } |\vec{v}| < \infty \text{ or } |\vec{v}| \leq \infty \qquad (6)$$

are observed as velocities of inertial systems.

Now consider a kinematical automorphism $i_1 \rightarrow i_2$ ("boost") whose matrix representation $f_{\vec{v}}$ is unique up to extensions of $A\,u\,t(P_2)$. Since $H_1 \cap H_2$ is mapped onto a subspace of H_2 we may arrange things such that $f_{\vec{v}}$ leaves $H_1 \cap H_2$ fixed. In the case $H_1 = H_2$ this already implies that $f_{\vec{v}}$ is a Galilei-transformation. In general, $f_{\vec{v}}$ is an element of

$$B \underset{def}{=} \{\underline{b} \in \underline{A\,u\,t}^E(k) \,\big|\, \underline{b}|H_1 \cap H_2 = 1\}. \qquad (7)$$

The latter can be shown to be an 1-dimensional Lie group, since its 1-component operates transitively and effectively on the 1-dimensional subspace of E_p generated by i_1 and i_2. If R is an appropriate rotation we have $\underline{R}\,f_{\vec{v}}\,\underline{R}^{-1} = f_{-\vec{v}}$, hence det $f_{\vec{v}} = \pm 1$.

A generating Lie algebra element for B thus may have the eigenvalues $\pm\lambda$, $\pm i\,\lambda$ or 0 as double value. In the first case the orbits of B are hyperbolas and their asymptotics. By symmetry arguments they must have the "right" position, such that B consists of 2-dimensional Lorentz-transformat-ions. Similarly the second case leads to 2-dimensional ro-tations and the third case, where only one eigenvector exists, to Galilei-transformations. In each case the full group $A\,u\,t(k)$ may be built up from a 1-parameter subgroup of boosts and extensions of spatial automorphisms. The group Aut(k) determines the kinematical structure since it allows to determine the hyperplanes H_i, $i \in J$, and its Euclidean structure. Hence we may conclude that the kinematical structure which we have researched is isomorphic to one of the standard kinematics considered in the introduction.

T13: If the kinematical structure k satisfies the axioms G, J, C, R, I, S and the considered geometry has three dimensions, then k is isomorphic to either the
    (i)   hyperbolic   or
    (ii) parabolic   or
    (iii) elliptic kinematics.

It seems that the precision of experiments with moving macroscopic bodies is not high enough to discriminate between these 3 possibilities. Even an elliptic mechanics may be possible, though a particle will then loose energy when accelerated. However, the existence of broad-casting is a strong evidence against elliptic Maxwell equations.

Fachbereich Physik der Universität 4500 Osnabrück,
Federal Republic of Germany

REFERENCES

Bourbaki, N.: 1968, Elements of Mathematics: Theory of Sets, Herman, Paris.
Brennich, H.: 1969, 'Zur Begründung der Lorentz-Gruppe von Süßmann', Z. Naturforsch. 24a, 1853-1854.
Ehlers, J., Pirani, F.A.E. and Schild, A.: 1972, Chapter 4 in L. O'Raiffeartaigh (ed.), General Relativity, Oxford.
Ehlers, J.: 1973, 'The Nature and Structure of Spacetime', in J. Mehra (ed.), The Physicist's Conception of Nature, Reidel, Dordrecht, pp. 71-91.
Freudenthal, H.: 1956, 'Neuere Fassung des Riemann-Helmholtz-Lieschen Raumproblems', Math. Z. 63, 374-405.
Freudenthal, H.: 1964, 'Das Helmholtz-Liesche Raumproblem bei indefiniter Metrik', Math. Ann. 156, 263-312.
Havas, P.: 1964, 'Four-Dimensional Formulations of Newtonian Mechanics and Their Relation to the Special and the General Theory of Relativity', Rev. Mod. Phys. 36, 938-965.
Janich, P.: 1969, Die Protophysik der Zeit, Bibliographisches Institut, Mannheim.

Janich, P.: 1980, Die Protophysik der Zeit, Suhrkamp,
Frankfurt am Main.

Kobayashi, S. and Nomizu, K.: 1963, Foundations of Differential Geometry, Interscience, New York.

Lévy-Leblond, J.-M.: 1965, 'Une nouvelle limite non-rela-
tiviste du groupe de Poincaré, Ann. Inst. Poincaré
A-III-1,1-12.

Ludwig, G.: 1974, Einführung in die Grundlagen der Theoretischen Physik, Band 1, Bertelsmann, Düsseldorf.

Mayr, D.: 1979, Zur konstruktiv-axiomatischen Charakteri-
sierung der Riemann-Helmholtz-Lieschen Raumgeometrien und
der Poincaré-Einstein-MinkowskischenÖRaumzeitgeometrien
durch das Prinzip der Reproduzierbarkeit, Doctoral dissertation, München.

Mayr, D.: 1981, Article in this volume.

Petrov, A.Z.: 1969, Einstein Spaces, Pergamon, Oxford.

Reichenbach, H.: 1928, 'Die Axiomatik der relativistischen
Raum-Zeit-Lehre', in A. Kamlah and M. Reichenbach (eds.),
Gesammelte Werke, Vieweg, Braunschweig (1979).

Robb, A.: 1936, Geometry of Time and Space, Cambridge.

Schmidt, H.J.: 1979, 'Axiomatic characterization of pyhsi-
cal geometry', Lecture Notes in Phys. 111, Springer,
New York.

Schutz, I.W.: 1973, 'Foundations of Special Relativity:
Kinematic Axioms for Minkowski Space-Time', Lecture
Notes in Math. 361, Springer, New York.

Süßmann, G.: 1969, 'Begründung der Lorentz-Gruppe allein
mit Symmetrie- und Relativitätsannahmen', Z. Naturforsch.
24a, 494-498.

Tits, J.: 1955, 'Sur certaines classes d'espaces 'homogènes
de groupe de Lie', Mém. Acad. Roy. Belg. Sci. 29 (3).

Dieter Mayr

A CONSTRUCTIVE-AXIOMATIC APPROACH TO PHYSICAL
SPACE AND SPACETIME GEOMETRIES OF CONSTANT CURVA-
TURE BY THE PRINCIPLE OF REPRODUCIBILITY

1. INTRODUCTION

The so-called space-problem, wellknown for more than hundred
years, is based on the famous works of B. Riemann, H.v.Helm-
holtz and S. Lie. It concerns questions about an axiomatic
characterization of physical space, precisely to the question
what topological and grouptheoretical assumptions may suffice to deduce the classification of Euclidean and non-Euclidean geometries of constant curvature. About fifty years
later a new complex of questions arose which offers much analogy to the space-problem, the axiomatic characterization of
the special relativistic spacetime – the so-called Poincaré-
Einstein-Minkowski spacetime-problem. Both sets of problems
have distinct issues in common, because both are invariance
problems of groups of automorphisms. In the space-problem
the invariance structures appear in the congruence or rigidity of spatial regions being conserved under displacements.
In the case of spacetime, the invariant structures enter into the Minkowskian metric and into the (Poincaré-invariant)
straight lines of events which are physically interpreted by
light rays and orbits of freely falling test particles.Both
invariance structures are closely interrelated; for instance,
spatial congruence und rigidity (rods) may be represented in
spacetime by orbits of freely falling particles (Ehlers 1973).
Thus, behind this conceptual interrelationship of rigidity
and inertia one may presume a combination of the space – and
spacetime-problem in which both appear as special cases of
a more general problem of invariance. At least this presump-
tion is supported by many correspondences and analogies between the Euclidean and Minkowskian geometry.
    The properties of invariance characterize, however, only
one half of the space- and of the spacetime-problem. Opera-
ting with hard objects one knows by experience that they may
occupy any location and position in space. Obviously a considerable richness of displacements is necessary for these

105

D. Mayr and G. Süssmann (eds.), Space, Time, and Mechanics, 105–123.

experiments. That this abundance of motions does not exist
a priori, is known at least since the discovery of Riemannian
manifolds in which these "mobility postulates" are not fulfilled in general. Against that, Euclidean and non-Euclidean
geometries (of constant curvature and sufficient isotropy)
are distinguished by free mobility, i.e. by the unrestricted
rotation and translation of rigid bodies (H. v. Helmholtz
1868). Thus, a geometrical characterization of space needs
the common effect of invariance and mobility. In spacetime
we meet an analogous connection which may be physically expressed by the principle of reproducibility of macroscopic
processes. In the following we shall illustrate more closely
this fundamental principle[1] and its significance for the
spacetime-problem.

If we imagine the continuance of a rod as a course of
events, we may thereby open a new prespective which goes
beyond the spatial invariance and mobility. The invariance
of the spatial extension of the rod is only the visual aspect
of a more general invariance property which is related to
spacetimelike entities - to courses (processes). Physically
one may describe this property in the following way. Under
certain precautionary conditions physical operations and
experiments are repeatable in the same way at new locations
and at other times such that all measurements and results
agree. The repetition of physically equivalent courses corresponds, geometrically, to a spacetimelike "displacement" (reproduction) of processes in a way that leaves unchanged any
part of the processes. This invariance property characterizes
reproducibility in much the same way as the relation of rigidity distinguishes the motions in space. Hence, a reproduction maps a process into a "displaced" process and both are
equivalent by the principle of reproducibility. Moreover,
idealized in the usual way the principle guarantees that reproductions are possible at all times and any positions, i.e.
anywhere in spacetime. Thus, reproductions may be represented by spacetime transformations. And if we use the concept
of reproducibility a sufficient richness ("free mobility")
of reproductions is ensured. Clearly, these idealizations
are the germs enforcing constant curvature.

The principle of reproducibility is not only an empirical fact, but also a basis of experience. For instance, the
relative gauging of rods and clocks leads always to the same
result, irrespective of their past history. Already 1918,
A. Einstein has called attention to this experimental fact.
From that one deduces traditionally the extension of terres-

trial concepts and laws to the universe. That is also one of
the conditions which makes it possible to do physics. Of
course, the extension is not without problems. If we diffe-
rentiate between micro-, macro- and astrophysical issues, it
is doubtful wether concepts and theories, developped in local laboratories, are compatible with experiments which dis-
close microscopic (quantum theory) and cosmic (gravitation
theory) spacetime domains. Hence, we meet this consistency
problem by the idealization of reproducibility to arbitrary
spacetime domains.

The analogy between the space- und spacetime-problem was
ignored amazingly long, before it was considered at the first
time by the mathematician H. Freudenthal (1964). In the beginning Freudenthal (1956) has brought the space-problem to
a certain completion following a preceding paper of J.Tits
(1955). Without assumption about differentiability and metric structures, i.e. without the usual concepts in all former treatments of the space-problem, the Lie-property of the
group was deduced exclusively by topological and grouptheoretical arguments. The deduction was essential based on a
theorem of Yamabe (1953). With Lie-algebraical methods Freudenthal simplified the main ideas of Tits' argumentation.
Later, in the combined space- and spacetime-problem, Freudenthal used the same mathematical tools. But there, he met a
situation considerably more complex as in the case of the
space-problem. Accordingly, he postulated the following presuppositions, before he specified some axioms proper.

Presupposition:

i)   A topological space X which is locally compact, connec-
     ted, separable and metrizable.
ii) A transitive group G of homeomorphisms which is, with
     the topological structure of continuous convergence,
     locally compact, separable and metrizable.
iii) A continuous realvalued function ϱ on X×X, with
     $\varrho(x,x) = 0$ and $\varrho(\tau x,\tau y) = \varrho(x,y)$ for all $x,y \in X$ and
     all $\tau \in G$ (quasi-metric).

Especially the last condition presents the combined space-
and spacetime-problem in a mathematical stage comparable with
the beginning of the space-problem. Moreover, Freudenthal's
basic concepts are genuin mathematically and topological abstract and therefore without direct interrelation to physical
concepts. What is, for instance, the physical meaning of the

second axiom of countability of a topological group? Of course,
here we meet the basic difference between mathematical and
physical theories - mathematical theories are far-reaching
independent from an interpretation of their basic concepts.
But if we consider the space- and spacetime-problem as physical problem, first of all, suitable basic concepts are necessary for an adequate physical interpretation.

Thus, it is our aim to establish an axiomatic theory in
which the presuppositions of Freudenthal's treatment may be
derived as theorems. Additionally, emphasizing the close
interrelationship of space and spacetime geometry, the axiomatic concepts and axiomatic structures ought to be related
to conceptions of G. Ludwig (1974) and H.-J. Schmidt (1979)
concerning the space-problem. Following these ideas we shall
determine our basic concepts in that domain of physics which
are accessible to simple experiments in a laboratory. Hence,
it must be operationally possible to relate straight forward the axiomatic structures to experience.

In general, however, the usual concepts, characterizing
space- and spacetime-geometry, like points, events, light
rays, particles (cp. Reichenbach (1977), Kronheimer-Penrose
(1967), Süßmann (1969), Ehlers et al. (1972), Schutz (1973)[2]),
are already abstract nouns which need more than a simple prephysical conception. Even if one shares the usual opinion
that event points may be represented by more and more smaller
courses, by "course-spots" as it were, nevertheless a precise
explication of the predicate "sufficiently small" is required
to deduce the point concept. Using straight lines, freely
falling particles, light cones, one runs into analogous
difficulties. The so-called "Protophysik", based on ideas of
H. Dingler (1928) , P. Lorenzen (1961,1978) and P. Janich
(1980), tries to eliminate the problems starting from manual
experienece and practice (which are not further explicated).
Following certain instructions and procedures "geometrical"
figures shall be constructed and used for measurement apparatus. These considerations, joined with some principles of
homogeneity, lead to some axiomatic ideas which are compared
with Euclidean geometry. And the desired Euclidean structure
is wanted to be a natural result of homogeneity and manual
experience. Consequently, the validity of geometry would be
reduced to the validity of the principle of homogeneity and
of other norms (with respect to manual instructions) - or
more general, physics would only be a theoretical stylizing
("Hochstilisierung" cf. Lorenzen (1978), p. 81) of manual
practice.

This position was criticized by various authors (cf. Böhme (1976), Süßmann (1979)). In our considerations we should like to use another idea of a physical axiomatization. Any physical theory contains certain sets, relations and functions, i.e. basic concepts which are interpreted straight forward or by means of precursor theories. Other terms of the theory are called "theoretical", because their experimental determination and interpretation will only succeed in principle within the complete theoretical framework. Hence, with respect to physical cognition one has to make efforts to select basic concepts and axioms (species of structures) in suche a way that a direct interpretation is possible for all terms. Consequently, a physical theory appears as an idealized, but yet immediate picture of its corresponding real facts and experiments - the domain of reality. This structuralistic point of view is due to the theory concept of G. Ludwig (1978) and may also be the leading idea for our analysis.

Now, the principle of reproducibility is based on the spacetimelike concepts of process and reproduction and offers thereby the basic concepts to our axiomatic approach. Accordingly, we will use the pretheoretical concepts of course (of events), part of a course and repetition of a course, to describe the real domain of physical spacetime geometry. In the mathematical theory they correspond to the basic concepts process, a relation $\sqsubset$ ("inclusion") and reproduction - or more precisely: a nonempty set P, whose elements are called processes, a binary relation $\sqsubset$ on P and a nonempty set $R \subset$ Pot (P×P), whose elements are called reproductions. Neither points, particles, freely falling particles or light rays (geodesics) nor spacelike, timelike or causal relations are used. In space geometry, the tripel (P, $\sqsubset$, R) may be easily reinterpreted by spatial regions, inclusion and displacement of spatial regions (cf. Schmidt 1979). On (P, $\sqsubset$, R) we postulate some axioms which may as well interpreted in space as in spacetime geometry. All axioms rely on empirical fact producable by simple experiments in a laboratory.

Finally, let us return to the mathematical aim of our considerations - the deduction of Freudenthal's presuppositions within the axiomatic (P, $\sqsubset$, R)-theory. After a connection with Freudenthal's treatise we obtain the classification of the Euclidian and non-Euclidian geometries of the Riemann-Helmholtz-Lie space-problem and of the pseudo-Euclidian and pseudo-non-Euclidian geometries of the Poincaré-Einstein-

Minkowski spacetime-problem.[3] Freudenthal (1964) calls his
approach a "reichlich unbefriedigenden Versuch, einen weiteren Rahmen zu finden, in den auch die pseudo-Euklidischen
und pseudo-nicht-Euklidischen Geometrien passen". The diffi-
culty was mainly based on how to generalize metrics which
are generated by indefinite quadratic forms. In our analysis
we try to do one step into this foundation problem.

2. PROCESSES AND REPRODUCTIONS - BASIC CONCEPTS

2.1 Courses and processes

The spacetimelike continuance of a body, or more general, the
spacetimelike existence of any object represents a course of
events. For our purposes it will suffice to restrict our
considerations to a certain class of physical operations in
a laboratory, to simple experiments carried out with bodies,
liquids, gases or physical fields. Rods in rest or in motion,
pendulums, chains or rolling balls are easy examples of
courses.[4] In the axiomatic theory courses correspond to (no
more specified) elements of a nonempty set P which are
called processes. Graphically, processes are represented by
certain domains of mathematical spacetime, e.g. by open relatively compact subsets of $\mathbb{R}^4$.

If we divide a course into several sections, or if we
restrict our observation only to some parts of the operational objects, in any case we obtain courses which are parts
of more comprehensive courses. Corresponding to this sub-
structure we introduce a binary relation "$\sqsubset$" on P. To
characterize some obvious properties of this inclusion relation, we postulate the relation to be reflexive, transitive
and properly ordered, i.e. $a \sqsubset b$ and $b \sqsubset a$ implies $a = b$
for all $a,b \in P$. Experiments with liquids or gases demon-
strate that different courses may include common subcourses.
These empirical structures are usually represented by the
introduction of an infimum $a \sqcap b$, or a supremum $a \sqcup b$ in analogy to the union of courses. As consequence we obtain the
structure of a lattice which is, by similar considerations,
distributive.

Axiom 1: $(P, \sqsubset)$ is a countably infinite and distributive
           lattice with a least element 0.

The postulated countability is due to the idealization usually
made by the actual presence of only finitely many members.

2.2. Repetitions and reproductions

Some more abstraction need the formal characterization of
process reproduction. In its idealized form the principle of
reproducibility postulates that, with respect to certain experimental precautions, courses are repeatable in an equivalent way, i.e. they agree in any experimental result. In
sufficient homogeneous spacetimes the equivalence is excellent-
ly corroborated by experiments. Making reproducibility more
familar, let us imagine an experiment (course) which is
equally performed on a fixed and on a moving (e.g. conveyor
belt) bases in a laboratory. It is a general fact of experience (Poincaré-Einstein principle of relativity) that, with
suitably chosen basis, the courses agree in all parts and
results. Clearly, the equivalence is the empirical foundation of the famous relativity principle. Later we shall
illustrate more experimental details.

Repetitions are practicable not only to single, but also to finitely many courses. By this procedure classes of
equivalent courses are obtained. In the axiomatic theory the
corresponding sets are extended to all processes such that
any repetition corresponds to a bijection (reproduction) on
P. Obviously, the global extension only makes sense, if a
homogeneous spacetime is assumed (global inertial systems).
In addition, we notice a further characteristic feature of
reproducibility. From relatively accelerated bases it is
wellknown that equivalently repeated courses are not possible
in principle. The difference appears in the fact that certain subcourses of the unrepeated course do not correspond
to any part of the repeated course. Consequently, an adequate
description of reproducibility requires the conservation of
the process lattice, i.e. reproductions have to be lattice
automorphisms. Finally, if we carry out repetitions of courses in the mentioned way, a successively performance (combination) of repetitions yields a repetition again, i.e. reproducibility induces a group structure on the lattice automorphisms. In the axiomatic theory, thus, we introduce a
nonempty set R ⊂ Pot (P×P) of reproductions and conclude
with the following

Axiom 2: R is a subgroup of the group of automorphisms of
$(P, \sqsubset)$.

3. PROCESSES AND REPRODUCTIONS – PROPERTIES

3.1. <u>Quasi-rigidity and coverings</u>

In general the Euclidian distance of events is not invariant under spacetime transformations, or in other terms, repro-ductions do not conserve the Euclidian form of processes. This is demonstrated by the following two examples of styli-zed Galilei $a \mapsto \tau a$ and Lorentz $b \mapsto \sigma b$ transformations:

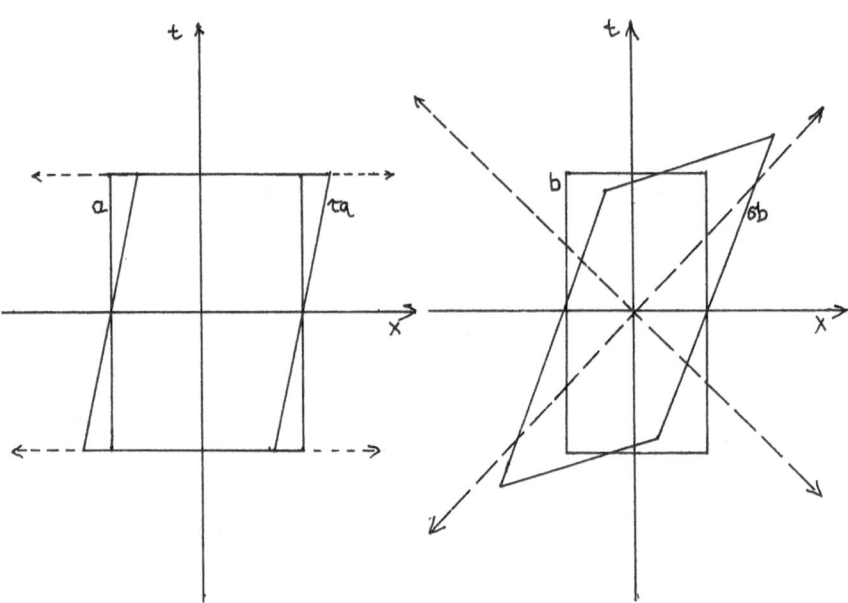

The broken lines point out the direction in which processes may be modified by boosts beyond all boundaries.[5] This is mathematically expressed by the noncompactness of the corres-ponding group (Galilei or Poincaré group), in constrast to the compact group of Euclidian rotations. Each group, how-ever, operates "rigidly" with respect to its proper concept of distance (quasi-rigidity). In the following we analyse more closely quasi-rigidity to characterize some particular structures of reproducibility.

For these purposes let us start with a simple physical experiment. Consider two equally constructed plates of glass on which some (n) lamps are fixed in equivalent positions respectively. The lights are so at the controls that each corresponding pair flashes synchronously. The first plate (G) rests on the table of the laboratory, the second (G') is moved parallel to G by a conveyor belt. Without auxiliary means an observer can easily make out, wether G' meets G, or in other terms, wether the two courses G and G' (represented by the spacetimelike progress of the plates) do intersect in spacetime. In particular, the analogous observation is possible for the corresponding lights L_i and L_i' which represent subcourses.$^{6)}$

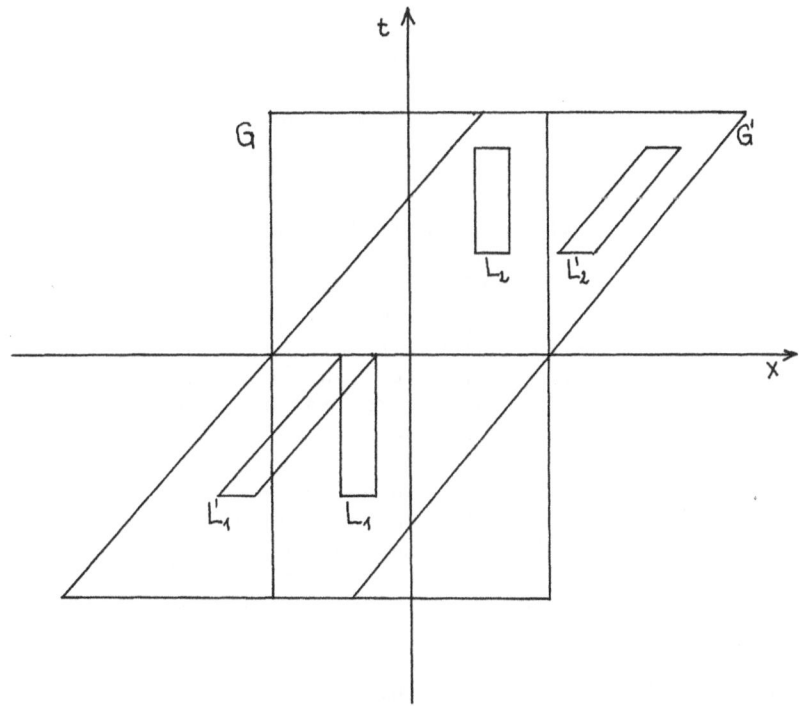

Various repetitions are performed by different velocities of the conveyor belt. In this way we get situations in which corresponding subcourses L_i and L_i' do or do not intersect. Clearly, the presence of intersections essentially depends

on the chosen repetition. The correlation between repetition
and intersection are due to an invariance structure described by the following empirical statement: It is always possible to select n subcourses L_i (sufficiently small) such that
the following condition is fulfilled for all repetitions: If
the corresponding subcourses do intersect $\overline{(L_i \cap L_i')} \neq \emptyset$,
i = 1,...n) all repeated subcourses L_i' remain inside the
original course G. This invariance structure is the essential feature of reproducibility. It is mathematically expressed by the fact that spacetime transformations operate
as rigidly, with respect to their proper geometry, as Euclidian transformation (quasi-rigidity). Thus, we may conclude
our considerations to the tentative statement $(P_o: = P/\{0\})$.

Quasi-rigidity:

$$\bigwedge_{a \in P_o} \bigvee_{n \in \mathbb{N}} \bigvee_{a_1,..,a_n \in P_o} \bigwedge_{\tau \in R} \bigwedge i = 1,..,n:$$

$$a_i \sqsubseteq a \wedge (\bigwedge j = 1,..,n: \tau a_j \cap a_j \neq 0 \rightarrow (\bigwedge k=1,..,n:$$

$$\tau a_k \sqsubseteq a)).$$

It is easily seen that the (spatial) rigidity axioms in
Freudenthal (1956) and in Schmidt (1979) are special cases
of quasi-rigidity (n=1). This connection reveals one of the
common roots of space and spacetime. Clearly, if we consider
the elements of P as spatial entities, quasi-rigidity implies the usual Euclidian rigidity. In the case of spacetime
quasi-rigidity determines a universe in which courses appear
as quasi-rigid objects, or in mathematical terms – using the
Lorentz metric, a process is rigid in the traditional sense.
    Now, let us stress a more technical point related to
quasi-rigidity. Obviously, the set $a_1,-,a_n$ of processes,
the so-called kernel of a, confine (by the intersection condition) the set of all reproductions to the so-called kernel
reproductions. Thus, one may expect that arbitrary small
kernels correspond to arbitrary small sets of kernel reproductions. This intuitive idea, however, is not quite ensured
by quasi-rigidity. To clear up this point we consider a
class of diffeomorphisms on $\mathbb{R}^n$ which coincide on a fixed
domain D with the identical transformation and which may act
in any way outside of D. Consequently, for any kernel
$d_1,-,d_n$ of D the given class of diffeomorphisms fulfills
quasi-rigidity, in contrast to the mentioned expectation. To
exclude those pathological cases we postulate the existence

of kernels which correspond to sets of kernel reproductions as small as we please.

Axiom 3:

$$\bigwedge a,b \in P_o \bigvee n \in \mathbb{N} \bigvee a_1,..,a_n \in P_o \bigwedge \tau \in R \bigwedge i=1,..,n:$$

$$a_i \sqsubset a \wedge (\bigwedge j=1,..,n: \tau a_j \sqcap a_j \neq 0 \rightarrow (\bigwedge k=1,..,n:$$

$$\tau a_k \sqsubset a \wedge \tau b \sqcap b \neq 0)).$$

If processes are represented in $\mathbb{R}^n$, the complete set of affine transformation is compatible with axiom 3, i.e. not only the usual spacetime transformations have to be taken into consideration, but also more general transformations which are combinations of dilatations and shearing trans- formations (i.e. all diffeomorphisms with constant strain) tensor). Later on we turn back to this problem.

Next, we define on R a topological structure by the sets

$$R(a): = \left\{ \tau \in R \mid \tau a \sqcap a \neq 0 \right\}, \quad a \in P_o$$

such that the set of finite intersections

$$\bigcap_{i=1}^{n} R(a_i), \quad n \in \mathbb{N}, \quad a_i \in P_o$$

forms an open neighborhood base of the identity in R. As consequence of axiom 3, R turns into a topological group. In the case of space geometry R(a) is compact, but in spacetime, this is obviously not true. In the mathematical part of the theory (cf. Mayr 1979) axiom 3 mainly implies the locally compact structure of the topological group (pre-Lie struc- ture).

In a simplified way one may experimentally copy the structure of the neighborhood base. Following the aforemen- tioned experiment repetitions are easily characterized by different velocities of the conveyor belt. With respect to the kernel condition $L_i \cap L_i' \neq \emptyset$, $i = 1,..,n$, a finite class of repetitions is obtained which corresponds to a certain neighborhood of the identity. Of course, the correspondence is established after a sufficient "completion" of the finite class, usually carried out by a suitably uniform structure (inaccuracy structures, cf. Ludwig 1978). In this sense axiom 3 may be experimentally confirmed in a laboratory.

In the following we should like to construct event points by the use of arbitrary "small" processes. For this purpose, first of all, we need an explication of the concept "small" or more precisely "small of a certain order". Corresponding to the idea of more and more smaller processes the concept of a filter seems to be appropriate for construction. A mere filter, however, is inadequate, because there may be finer filters which contradict the point concept. On the other hand, if we use so-called ultra filters (finest filters), the resulting set of point does not carry the natural topology, whose neighborhoods are induced by the elements (processes) of the filters. Accordingly, one needs a concept of a filter which stands in the middle, as it were, i.e. a filter which generates points and which conserves the natural concept of neighborhood. From topology it is wellknown that so-called Cauchy filters exactly fulfill these conditions. Of course, their definition presupposes uniform structures, or roughly, a globally uniform criterion of comparison. [7]) Using the neighborhood base of the identity we may generate these structures such that any process is characterized by its spacetimelike "size". A tentative definition would read (U is an element of the neighborhood basis $\mathcal{B}$ of the identity):

Definition:

$a \in P_o$ is called covering-small of order U, $cs(a,U)$, iff

$$\bigwedge_{b \in P_o} \bigvee_{n \in \mathbb{N}} \bigvee_{\tau_1,..,\tau_n \in U} : b \sqsubset a \rightarrow a \sqsubset \bigsqcup_{j=1}^{n} \tau_j b.$$

For sake of technical simplicity we shall often use the following more general characterization.

Preuniformity:

$a \in P_o$ is called U-small, $sm(a,U)$, iff

$$\bigwedge_{b,c \in P_o} \bigvee_{\tau \in U} : b,c \sqsubset a \rightarrow \tau b \sqcap c \neq 0.$$

The designation "preuniformity" may immediately be justified in connection with axiom 4. In analogy to uniform spaces, we should like to define so-called Cauchy prefilters by arbitrary U-small processes. Their existence needs the following covering postulate.

Axiom 4:

$$\bigwedge_{a,b\in P_o} \bigwedge_{U\in B} \bigvee_{b'\sqsubset b} \bigvee_{n\in \mathbb{N}} \bigvee_{\tau_1,..,\ \tau_n \in R} \bigwedge_{i=1,..,n}:$$

$$cs(\tau_i b',U) \wedge a \sqsubset \bigsqcup_{j=1}^{n} \tau_j b'.$$

Axiom 4 guarantees that for any process, there is a finite covering by U-small processes, i.e. a kind of local precompactness.

Clearly, the quantification over the elements of B ensures the existence of arbitrary small processes. Consequently, there are Cauchy prefilters on P and what is more, any Cauchy prefilter includes a unique, minimal (coarsest) Cauchy prefilter. As usual, points may be identified with these minimal filters, much in the same way as one deals with completion of the rationals to the reals. Points, constructed in this way, represent events in spacetime.

On the set E of points the relation of preuniformity induces a uniform structure; its entourages are generated by

$$N_u := \left\{ (x,y) \in E \times E \ \middle| \ \bigvee_{a\in P_o}: a \in x,y \ \wedge \ sm(a,U) \right\}, \ U \ B.$$

$a \in x,y$ reads that the points x and y, represented by minimal Cauchy prefilters, include the process a. With these structures E is a complete Hausdorff space and its countable base of entourages implies the metrizibility of E. A process $a \in P_o$ corresponds to an open relatively compact subset $\tilde{a}$ of E, i.e. the relations $a \in x$ and $x \in \tilde{a}$ are equivalent. Moreover, reproductions map minimal filters into minimal filters such that any $\tau \in R$ corresponds to a uniform automorphism $\tilde{\tau}$ on E, defined by $x \longmapsto \tilde{\tau}x := \{\tau a \mid a \in x\}$. Thus, R canonically generates a topological group $\overset{*}{R}$ of uniform automorphisms on E. If we complete $\overset{*}{R}$ with respect to its two-sided uniformity (cp. Bourbaki (1966), III, § 3.4), we obtain a locally compact group $\overset{*}{R}$ which is separable and metrizable.

3.2. Process chains

To deduce Freudenthal's presuppositions (cp. section 1) we have to prove the connection of the space E. This is usually done by chains, or in our terms, by chains of processes connecting any two other processes. Process chains are experimentally constructed by roughly periodic courses, for instance by pendulums, which are transported form one process to another.

Definition: The set $\{c_1,..., c_n\} \subset P_o$ is called a process chain between a and b of P_o, $\langle a \,|\, c_1,...,c_n |\, b \rangle$, iff

$$\bigwedge i = 1,..,n-1: \; a \sqcap a_1 \neq 0 \wedge a_i \sqcap a_{i+1} \neq 0 \wedge a_n \sqcap b \neq 0.$$

If the existence of process chains should imply the connection of E, we first have to guarantee connected chains, or in other terms, we have to postulate processes connected in the usual topological sense. We formulate this idea without topological concepts (preconnection).

Definition: $a \in P_o$ is called chainconnex, ch(a), iff

$$\bigwedge b,b' \sqsubset a \bigwedge U \in B \bigvee n \in \mathbb{N} \bigvee c_1,..,c_n \in P_o \bigwedge i,j = 1,..,n:$$

$$c_i \sqsubset a \wedge sm(c_j,U) \wedge \; \langle b \,|\, c_1,...,c_n |\, b' \rangle \;.$$

Now, the existence of chainconnex process chains implies the connection of E. Moreover, any chain link of a chainconnex chain is again chainconnex. Hence, if two processes are connected by a chainconnex chain, they include chainconnex subprocesses which belong to the chain. This structure implies the local connection of E.

Axiom 5:

$$\bigwedge a,b \in P_o \bigvee a' \in P_o, \; a' \sqsubset a \bigvee n \in \mathbb{N} \bigvee c_1,...,c_n \in P_o:$$

$$ch(a') \wedge ch(\bigsqcup_{i=1}^{n} c_i) \wedge \langle a \,|\, c_1,...,c_n |\, b \rangle \;.$$

From the covering condition in axiom 4 it is easily deduced that the group $\hat{R}$ operates transitively on compact subset of E. Following axiom 5 any two points may be connected by chains which are relatively compact subsets. This fact implies the transitivity of $\hat{R}$.

3.3. Quasi-metric

It remains to prove the existence of a continuous function

$$\rho: E \times E \to \mathbb{R}, \qquad \rho(x,x) = 0$$

which is invariant with respect to transformation of $\hat{R}$ (quasi-metric). Clearly, in particular cases, ρ represents

the Euclidian or Lorentzian metric. To construct a quasi-
metric we use the ordered structures of some hypersurfaces
which are orbits of the stability group of x

$$J_x: = \left\{ \tau \in \hat{R} \mid \tau x = x \right\}.$$

In space or spacetime the orbits are spherical or hyperbolic
surfaces, respectively, defined by

$$J_x^o(y): = \left\{ \tau y \mid \tau \in J_x^o \right\}, \qquad y \in E;$$

J_x^o is the identity component of J_x. The orbits are often
called spheres of constant distance. In the case of null
distance we shall call the orbit "isotropy cone". In space-
time it is represented by the light cone (null cone); in
space it reduces to the fixed point x. Using the concepts of
process and reproduction to define the isotropy cone, we re-
call that a reproduction modifies a process in directions
characterizing the light cone (cp. sect. 3.1). Accordingly,
the union of all reproduced processes yields a (hyperbolical)
neighborhood

$$J_x^o(a): = \bigcup_{\tau \in J_x^o} \tau \tilde{a}$$

of the isotropy cone and, thus, an obvious definition would
read:

$$I_x: = \overline{\bigcap_{a \in X} J_x^o(\tilde{a})}.$$

Consequently, the isotropy cone is not empty, closed and
connected. And it is invariant with respect to $\hat{R}$, i.e.
$\tau I_x = I_{\tau x}$ for all $\tau \in R$.

Special relativistic spacetime is separated by the light
cone into disjoint parts (components) in which the orbits
are linearly ordered. To obtain these ordered structure, we
introduce a concept of separation. Following J. Tits (1955)
we define a separation between three subsets A,B,C of E.

Definition: B separates A from C, S(A,B,C), iff there are
two nonempty open subsets P_1, P_2 of E such that
(i) $A \subset B \cap P_1$ and $C \subset B \cap P_2$;
(ii) $P_1 \cap P_2 \subset B$;
(iii) $B \cup P_1 \cup P_2 = E$.

In Tits' argumentation separation serves to prove the linear

order of the orbits (spherical shells in the case of space
geometry). He mainly uses the postulated property that space
transformations operate rigidly, i.e. they are isometries. In
our case this is not true. The problem is related to the
following fact. The uniform structure of E is represented by
the usual uniformity of $\mathbb{R}^n$, in space just as in spacetime.
But spacetime transformations do not operate rigidly, they
deform the Euclidian distance (uniform automorphisms) and
Tits' argumentation fails. To prevail these problems we define a stronger relation.

Definition: A and C are strongly separated by B, $T(A,B,C)$, iff
(i) $S(A,\bar{B},C)$;
(ii) $P_1 \cap P_2 = \emptyset$ and $P_i \cap \overset{\circ}{B} = \emptyset$, $i = 1,2$.

$\overset{\circ}{B}$ is the topological interior of B. If we replace B by an   8)
orbit, the space is separated in two open, disjoint subsets.
Moreover, T is only true, if the orbit is closed and nowhere
dense. Now, we are prepared to take over Tits's axiom about
the ordered structure of orbits - a further hint to the surprisingly close connection of space and spacetime geometry.

Axiom 6:

$$\bigwedge x,y,z \in E: (y \notin I_x \wedge z \in C_x(y)) \longrightarrow$$

$$\longrightarrow (T(\{x\}, J_x^o(y), J_x^o(z)) \vee T(\{x\}, J_x^o(z), J_x^o(y))).$$

$C_x(y)$ is the topological component of y in the subspace E/I_x.
Axiom 6 enforces the exclusion of dilatations just as the
Galilean spacetime (shearing transformations).
    The set of orbits

$$0_x^o(y) : = \left\{ J_x^o(z) \mid z \in C_x(y) \right\}$$

is invariant with respect to $\widehat{R}$ and linearly ordered by the
relation T. Moreover, the set $0_x^o(y)$ is open in the quotient topology of the orbital equivalence relation, continuous in the sense of Dedekind (cf. Hausdorff 1949) and it
includes a countably dense subset. Consequently, $0_x^o(y)$ is
isomorphic to an open interval of the reals and we may easily define a function with the desired properties of the quasi-metric. Hence, we have proved the mentioned presuppositions (section 1) and obtain with Freudenthal's treatment
the classification of $(E, \varsigma, \widehat{R})$-geometries:

1) n-dimensional Euclidian and non-Euclidian geometries
   (n $\geq$ 2), i.e. parabolic and hyperbolic, elliptic and
   sperical spaces.
2) n-dimensional pseudo-Euclidian and pseudo-non-Euclidian
   geometries (n $\geq$ 3), i.e. pseudo-Riemannian manifolds of
   constant curvature (Minkowski and de Sitter spaces).

NOTES

1) Indeed, the principle is of fundamental importance. On the
   one hand, it is in its abbreviated form "under equal conditions equals happens" equivalent to the deterministic
   reinforcement of the causality principle - "Equal causes
   imply equal effects". On the other hand, the principle of
   reproducibility is the base on which the principle of relativity, or in a local view, the (semistrong) principle
   of equivalence is founded.
2) Every axiomatic approach to spacetime uses anywhere the
   principle of reproducibility. Suprisingly often, however,
   it is stated neither explicitely nor even axiomatically.
3) The general character of the mainly local $(P, \sqsubset, R)$-axioms
   suggests a more general spacetime-problem. At present the
   author is working on an analogous approach to general relativistic spacetime based on a suitable generalization of
   reproducibility (local reproducibility).
4) For more perception let us illustrate a "constructive" representation of courses. Consider a flat disc on a table,
   photographed n-times by a fixed camera. After development
   all negatives coincide. Now, if we put the negatives one
   upon another (like playing-cards), the discs form a cylinder which is nothing else than a material picture of the
   observed "disc course". The height of the cylinder corresponds to the time of observation. Other "figurative"
   courses are easily obtained by moving cameras.
5) These directional structures may be used to define the so-
   called isotropy cone. Graphically it is the set of all
   points with null distance. In spacetime the isotropy cone
   coincides with the light cone.
6) The preference of a Galilei transformation, represented in
   the following figure, is not essential to our considerations.
7) One should beware of seemingly local ideas, as for example
   concentric, more and more smaller balls. Because in those
   examples one uses, more or less unnoticed, metric or uniform, i.e. global concepts of comparison.

8) Separation by orbits is also postulated in Freudenthal
(1956). Amazingly enough, separation is, in some sense,
as effective as distance concepts, postulated in former
axiomatization of the space-problem. Among other things
separability enforces the Lie property of the group.

REFERENCES

Böhme, G. (ed.): 1976, "Protophysik". Suhrkamp, Frankfurt.
Bourbaki, N.: 1966, "Elements of Mathematics, General Topo-
logy". Herman, Paris.
Dingler, H.: 1928, "Das Experiment. Sein Wesen und seine Ge-
schichte", Reinhard, München.
Ehlers, J.: 1973, "Survey of General Relativity Theory", in
W. Israel (ed.), Relativity, Astrophysics and Cosmology,
Reidel, Dordrecht.
Einstein, A.: 1905, "Zur Elektrodynamik bewegter Körper",
Ann.d.Phys. 17, 891.
Ehlers, J., F.A.E. Pirani and A. Schild: 1972, in L.O'Rai-
feartaigh (ed.), "General Relativity", Clarendon Press,
Oxford.
Freudenthal, H.: 1956, "Neuere Fassung des Riemann-Helm-
holtz-Lieschen Raumproblems", Math. Zeitschr. 63, 374 -405.
Freudenthal, H.: 1964, "Das Helmholtz-Liesche-Raumproblem
bei indefiniter Metrik", Math. Annalen 156, 263-312.
Hausdorff, F.: 1949, "Grundzüge der Mengenlehre", Leipzig
(1914) Chelsea P.C. New York.
Helmholtz, H.v: 1868, Nachr. Ges. Wiss. Göttingen, 193-221.
Kronheimer, E. and R. Penrose: 1967, "On the Structure of
Causal Spaces", Proc.Camb.Phil.Soc.63, 481-501.
Lie, S.: 1890, "Über die Grundlagen der Geometrie", Berich-
te Ges.Wiss. Leipzig 42, 284-321, 355-418.
Lorenzen, P.: 1960, "Entstehung der exakten Wissenschaften",
Springer, Berlin.
Lorenzen, P.: 1978, "Theorie der technischen und politischen
Vernunft", Reklam, Stuttgart.
Ludwig, G.: 1974, "Einführung in die Grundlagen der theore-
tischen Physik", Bd. 1: Raum, Zeit, Mechanik; Vieweg,
Braunschweig.
Ludwig, G.: 1978, "Die Grundstrukturen einer physikalischen
Theorie"; Springer, Berlin.
Mayr, D.: 1979, "Zur konstruktiv-axiomatischen Charakteri-
sierung der Riemann-Helmholtz-Lieschen Raumgeometrien und
der Poincaré-Einstein-Minkowskischen Raumzeitgeometrien
durch das Prinzip der Reproduzierbarkeit", Dissertation,

University of Munich.

Minkowski, H.: 1915, "Das Relativitätsprinzip", Jahresber.
d.Deutsch.Math. Ver. 24, 372.

Poincaré, H.: 1905, "Sur la dynamique de l'électron", Paris
C.R. 140.

Reichenbach, H.: 1977, "Philosophie der Raum-Zeit-Lehre",
Bd. 2 Gesammelte Werke, A. Kamlah, M. Reichenbach (eds.)
Vieweg, Braunschweig.

Riemann, B.: Habilitationsvortrag (1854), 1867 published in
Abh.Ges.Wiss. Göttingen 15.

Schmidt, H.-J.: 1979, "Axiomatic Characterization of Physical Geometry". Lecture Notes in Physics, 111.

Schutz, J.W.: 1973, "Foundations of Special Relativity: Kinematic Axioms for Minkowski Space-Time", Notes in Mathematics, Springer.

Süßmann, G.: 1969, "Begründung der Lorentz-Gruppe allein mit
Symmetrie- und Relativitätsannahmen", Z.Naturf. 24a,
495-498.

Süßmann, G.: 1979, "Kennzeichnung der Räume konstanter Krümmung", in Materialienhefte des Schwerpunkts Mathematisierung, Bielefeld.

Tits, J.: 1955, "Sur certaines classes d'espaces homogènes
des groupes de Lie", Mém.Acad.Roy. Belg. Sci. 29 (3).

Yamabe, H.: 1953, "A Generalization of a Theorem of Gleason",
Ann. of Math. (2) 58, 351-365.

Erhard Scheibe

INVARIANCE AND COVARIANCE [1]

According to its title the present paper has a twofold aim.
As far as invariance is concerned I want to argue that the
numerous invariances of physical laws can essentially be
reduced to one concept of invariance that is already part
of the concept of kinds of mathematical structures that are
used to formulate physical theories. The second aim of the
paper is to present the concept of covariance not as any
kind of invariance but rather as a concept of equivalence
between two formulations of a physical theory that are already invariant, one of which, however, has a higher 'degree'
of invariance than the other. As regards the actual use of
the terms 'invariance' and 'covariance' in physics and the
occasional discussions of it in philosophy of physics I do
not want to entertain any criticism of the literature
although the somewhat deplorable state of affairs concerning
the two concepts would justify this. The present paper will
thus be selfcontained, and I will pay attention to what
usually is meant by invariance and covariance only to the
extent that makes it sufficiently clear what I am talking
about.

To give a brief outline of my argumentation I begin
with calling attention to the fact (part I) that one of our
fundamental metamathematical concepts, the concept of a
species of structures, is essentially defined by a condition
of invariance. On the other hand, the mathematical part of
every physical theory (the formalism that is peculiar to it)
presumably can be reformulated as a species of structures.
Consequently, to the extent to which this is true an invariance condition would be built into a physical theory from
the very outset. This naturally leads to the question what
this kind of invariance, canonical invariance as it will be
called, has to do with the well known invariances appearing
in physics as characteristic properties of its fundamental
laws. One proper context in which this question can be discussed is given by those physical theories whose mathematical formulation is based on species of structures called
manifolds (part II). Most important examples are theories,

125

D. Mayr and G. Süssmann (eds.), Space, Time, and Mechanics, 125–147.
Copyright © 1983 by D. Reidel Publishing Company.

e.g. Einstein's theory of gravitation, that are founded on
a theory of spacetime as the basic manifold. But also the
configuration spaces and phase spaces of classical mechanics
are cases in point. For a large class of these theories
(based on manifolds) it can be proved that canonical invariance is essentially equivalent to the common invariances
of physical laws, and it seems plausible that complete generality can be achieved. As a consequence, no invariance requirements beyond the one that is already part of the definition of a species of structures would have to be imposed on
a physical theory. This result can be viewed as a substanti-
ation and justification of what is nowadays sometimes recommended as the 'coordinatefree method' in physics.

  Transcending the concept of invariance the concept of
covariance (part III) points to the fact that two species
of structures can be equivalent and as such can become the
formalism of one and the same physical theory although their
degrees of invariance are different. They are, however,
supposed to be comparable in the sense that the groups underlying the invariances are related by inclusion. Covariance
as equivalence takes up the old issue that came up with the
general theory of relativity to increase at will the invariance of a physical law. Maxwell's equations in their lorentz
invariant form, for instance, can be <u>made</u> <u>invariant</u> under
arbitrary differentiable transformations by explicitly introducing the Minkowski metric as a tensor field. If this is
done two new facts occur: the invariance of the new equations
and their equivalence to the equations from which one started. But the new equations are invariant under the larger
group in exactly the same sense as were the original equations
with respect to the smaller group. Therefore, if there is
any need for a new concept then it is with respect not to
invariance but to equivalence.

  This can be shown more convincingly in the context of
the socalled 'principle of general covariance'. If this
principle is taken to mean that physical laws should be
invariant under arbitrary differentiable coordinate transformations then, as is well known and will here be confirmed
anew, it can be trivially satisfied unless further requirements are included. With respect to the converse relation
of covariance, however, the situation is altogether different: The reduction of the group of coordinate transformations is not always possible, and if it is possible this
may very well be a nontrivial result. In applying this
relation of reduction I shall argue that the general theory

of relativity has to be estimated on account of its irreducibility and <u>not</u> as is frequently done with regard to the possibility of finding covariant versions of other physical theories on the invariance level of general relativity.

The three concepts of species of structures, of the
invariance inherent in them and of the equivalence between
them are due to Bourbaki (1968, Ch. IV). They are, however,
founded on a particular formulation of logic and set theory
that I do not want to take over. I will rather assume a modern formulation of the system of Zermelo-Fraenkel (ZF). This
means that set theory is presented as a standard theory (cf.
Shoenfield 1967). For convenience I will even work with an
extension by definitions of ZF that is rich enough to contain
all the auxiliary concepts that will constantly be needed
(ZFT). The reader may or may not assume that ZFT (and hence
ZF) is given an interpretation in the modeltheoretical sense.
Such an assumption would at least not be necessary since all
metatheoretical concepts that will be introduced will concern
only syntactical entities (formulas or terms).

1. SPECIES OF STRUCTURES

Roughly spoken a species of structures is an <u>axiom</u> <u>system</u>
of a special kind: There is a subdivision of its concepts into
basic concepts $X_1, \ldots, X_n$ and typified concepts $s_1, \ldots, s_m$
as well as a corresponding subdivision of the axioms into
proper axioms α_i and typifications $\gamma_1, \ldots, \gamma_m$ such that
1) γ_μ determines the type of s_μ relative to the X_ν, i. e. it
determines the 'nature' of the entities falling under s_μ
relative to those falling under any of the X_ν, whereas
2) the γ_μ by themselves and the α_i by <u>fiat</u> do <u>not</u> determine
the 'nature' of the entities falling under the X_ν relative
to each other or to any concept 'presupposed' by the axioms.

To give the details in settheoretical terms (cf. Bourbaki
1968, Ch. IV) we think of an axiom system as being a finite
set of formulas of ZFT depending on certain variables indi-
cating the sets ('concepts') that are interrelated by the
axioms. We then have their subdivision into the <u>base</u> <u>sets</u> X
(short for: $X_1, \ldots, X_n$) and the <u>typified</u> <u>sets</u> s (short for:
$s_1, \ldots, s_m$). The <u>typifications</u> γ_μ of 1) are formulas

$$s_\mu \in \sigma_\mu(X) \tag{1}$$

where σ_μ is a scale term, i. e. a term constructed from its
arguments by successively applying one of the operations that

yield a power set (Pot) or a cartesian product ($\times$). In the
case before us the arguments are the X and possibly further
constant (!) terms that are available in ZFT. To keep the
notation simple the latter will not made explicit but must
be kept in mind. The 'nature' of the s (or – as above – of
their elements) is determined by (1) relative to the X and
constant sets from ZFT in the sense that the s are elements
of the power set of ..... the cartesian product of ..... the
X and the constants. The remarkable thing about (1) is that
it provides counterparts of all the predicates and terms of
arbitrary arity and order as they appear in the various
independent logical calculi. In the present context a scale
term σ is called a type of structures and a system of sets
X and s satisfying (1) a structure of type σ.

Having done with the determination of the s with respect
to the X and constant sets announced in 1) we have now to
prepare the ground for the invariance condition taking charge
of 2). To do this we define the canonical σ- representation
to be the following assignment: Given a scale term σ and
bijections f from sets X onto sets X' (i. e. according to
our convention bijections f_ν from X_ν to X_ν') a bijection f^σ
from $\sigma(X)$ onto $\sigma(X')$ is uniquely determined by the two
recursive conditions that for the power set operation

$$f^{\mathrm{Pot}\,\sigma}(s') = \{f^\sigma(s) \mid s \in s' \wedge s \in \mathrm{Pot}\,\sigma(X)\} \quad (2a)$$

and for the cartesian product operation

$$f^{\sigma_1 \times \sigma_2}(\langle s_1, s_2 \rangle) = \langle f^{\sigma_1}(s_1), f^{\sigma_2}(s_2) \rangle \quad (2b)$$

We go on to define f to be a σ - isomorphism from $\langle X, s \rangle$ onto
$\langle X', s' \rangle$ if and only if $\langle X, s \rangle$ and $\langle X', s' \rangle$ are structures of
type σ and the f are bijections from the X onto the X'
such that $f^\sigma(s) = s'$. It then follows immediately for the
typifications (1) that

$$iso_\sigma(X, s; X', s'; f) \rightarrow s \in \sigma(X) \leftrightarrow s' \in \sigma(X') \quad (3a)$$

is provable (in ZFT). By analogy we now require that the
proper axioms are canonically σ- invariant in the sense that
for their conjunction α

$$iso_\sigma(X, s; X', s'; f) \rightarrow \alpha(X; s) \leftrightarrow \alpha(X'; s') \quad (3b)$$

is provable (in ZFT).

An axiom system consisting of typifications (1) and proper axioms satisfying (3b) is called a <u>species of structures</u>. In Bourbaki (1968, Ch. IV) the crucial property (3b) of α is called 'transportability'. By using the term 'canonical invariance' instead I want to emphasize that, if anything, then (3) is a condition of invariance. The intuitive idea standing behind any definition of invariance is the idea of something that remains unaltered while something else on which it depends is changed. In the present case what remains the same is, expressed in semantical terms, the truth value of a proposition, and what is changed is that about which the proposition is a proposition: If the structure $\langle X, s \rangle$ is replaced by a structure $\langle X', s' \rangle$ isomorphic to the first then α is true about the second if it is true about the first and false about the second if false about the first. Extreme cases of invariance of truth values would be given by formulas provable or refutable in ZFT as, for instance, (3a) or (3b) themselves. In a model of ZFT such formulas would be true or false no matter what interpretations their variables are given, and this means that any replacement of the values of these interpretations would leave truth values invariant. This complete freedom of replacement is restricted in (3) in two ways: The base sets X may only be replaced by sets resulting from them by bijections f, and the typified sets s only by their images under the canonical representations of the f. This makes room for many more formulas becoming invariant in the sense of (3). On the other hand, the restrictions are still wide enough to prevent the axioms from saying anything about 'nature' of the X in the sense that their elements were determined relatively to each other or to any constant set. Without bringing this idea under precise terms let us imagine some clear cases in which one would say that such a determination occurs. Cases in point would be given by all typifications (1) with one of the base sets instead of s_μ and the rest of the base sets instead of X as our axiom α. It is evident that such a relation can be destroyed by isomorphisms and therefore would contradict (3b). The same would be true for axioms α saying that one base set X has an empty (or nonempty) intersection with another base set or constant set. Thus even such negative determinations are ruled out by our invariance condition.

Examples for species of structures abound from mathematics: In point of fact all the well known concepts of a group, ring, vector space, topological space, manifold, fibre bundle etc. are defined by axioms that can easily be reconstructed

as so many typifications (1) and axioms proper satisfying
(3b). It is likewise a fact that these mathematical concepts
are frequently applied in theoretical physics, and at least
in the mathematical treatment of higher level theories such
as quantum theory and the theory of general relativity the
use of species of structures has been generally acknowledged:
'A general lesson to be drawn from the development of the
theory of relativity is that it is desirable to analyse in
detail the various structures inherent in the mathematical
models used to describe physical phenomena' (Trautman 1972,
p. 85). The multifarious application of species of structures
in theoretical physics does, however, not mean that the
common conceptual basis of this mathematical field is sufficiently understood. By now there exists only one thorough-
going attempt of a general conceptual reconstruction of
physics on the basis (as far as mathematics is concerned) of
the concept of species of structures (Ludwig 1978). There
is, admittedly, also the set theoretical approach of Suppes,
Sneed and others, and it has recently been claimed that this
approach could be viewed as an extension of the Bourbaki programme to science (Stegmüller 1979). Apart from several objections that could be made against this claim the set theoretical approach can be left out of consideration here because it does never focus on the defining conditions of a
species of structures which, however, make all the difference
with regard to the goal of this paper.

   As I have tried to argue in a previous paper (Scheibe
1978) species of structures occur in physical theories not
only as incidental concomitants. Rather they can be used to
characterize a physical theory as a whole as far as this is
possible without regard to their empirical interpretation.
In particular this characterization includes physical laws
in the narrower sense and with them the sort of things in
whose invariance properties we are primarily interested.
However, the characterization can not be achieved solely by
the simple and general species of structures that are the
building blocks of modern mathematics. Rather we must have
recourse to richer species built up from the simpler ones:
'As a rule, rich structures are used in physics; those of
differential manifolds carrying additional geometric objects
and of Hilbert spaces with preferred sets of operators'
(Trautman 1972, p. 85).

   A general concept particularly helpful in making clear
what is at issue here is the concept of an extension of a
species of structures. Given a species (σ_0; α_0) it can be

extended by adding a (not necessarily) new typification $s \in \sigma(X)$ and axiom $\alpha(X; s_0, s)$, both referring to the old base sets, such that $(\sigma_b \times \sigma; \alpha_0 \wedge \alpha)$ again is a species of structures. It follows from the assumption that $(\sigma_b; \alpha_0)$ already is a species of structures that the canonical $(\sigma_b \times \sigma)$ - invariance of $\alpha_b \wedge \alpha$ is equivalent to the provability of

$$iso_{\sigma_0 \times \sigma}(X; s_0, s; X'; s_0', s'; f) \wedge \alpha_0(X; s_0)$$
$$\Rightarrow \alpha(X; s_0, s) \leftrightarrow \alpha(X'; s_0', s') \tag{4}$$

In this way the 'invariance part' of (3b) can be isolated for the underlined axiom α. Now, in physical practise one frequently proceeds from axioms that in a systematic account of the species of structures involved would be added only at the very end. It is this methodological inversion that can be elucidated by the concept of extension.

Take ordinary quantum mechanics as an example. The core of it, the physical law that really matters, is Schrödinger's equation (with $\hbar = 1$)

$$i \dot{\psi} = H \psi \tag{5a}$$

Asking for the axiomatic background of this equation it turns out that it is 'only' the last step in a series of extensions The series would start with an abelian group X with addition A which then would be extended into a complex vector space by adding a scalar multiplication B. The next extension would lead to a Hilbert space by adding a metric C, and this is the point where the structural aspect usually comes to a stop. But in principal the extension could be continued with a self-adjoint linear operator H and an X-valued and $\mathbb{R}$-argument function ψ as two new structures that are related by (5a). The only requirement would be that the whole axiom that we have obtained is canonically invariant with respect to the total typification of the five structures imposed on X, and this is indeed the case. In this way not only is the crucial equation (5a) submitted to a species of structures. It also has received the 'right' invariance property: The invariance of (5a) under the transformations

$$\psi'(t) = U \psi(t), \qquad H' = U H U^{-1} \tag{5b}$$

with a unitary transformation U of X that is usually considered is but a special case of canonical invariance. It

occurs if in the canonical representation of U not only X
but also A, B and C are left invariant which is but another
way of saying that U is unitary. All this comes out of (4) if
α_0 refers to the Hilbert space and α to the properties of H
and ψ including (5a) and if, finally, $X' = X$ and $S_0' = S_0$.
Then $s' = f(s)$ turns out to be (5b), and the conclusion of
(4) is the restricted invariance usually considered.

2. INVARIANCE

The foregoing example is a particularly favourable case to
my point that the invariance properties of physical laws can
be reduced to the canonical invariance in species of structures. The reason is that general quantum mechanics is used
to be presented at a considerable high level of abstraction
which sometimes comes very close to its presentation as a
species of structures. Actually the situation is not altogether different in the field that I am going to enter now:
Manifolds of this or that kind are frequently introduced in
a manner that makes it not too difficult to reformulate the
definitions in terms of species of structures. Sometimes the
abstract presentation even is highly recommended as a 'coordinate free' or 'intrinsicly invariant' method. But seldom, if
ever, is it precisely stated in what sense we get rid of
coordinates and by what features the invariances so characteristic of the old coordinate dependent formulations are replaced.
    In order to clear up the matter I start this investiga-
tion with a particularly simple class of species of manifolds
which in turn will be introduced in two steps. In the first
step species of global coordinate manifolds (gcm) are defined.
In a gcm the intuitive idea of a set F of preferred coordinate
systems used to label all (!) elements of an otherwise arbitrary set M is given a precise formulation. The species of
gcm are parametrized by two external parameters (i. e. constant terms being available in ZFT): a natural number n and
a group G of homeomorphisms of the n-dimensional real number
space $\mathbb{R}^n$ with its usual topology. Typification and axiom
defining the species gcm (n,G) are

$$F \in Pot^2(M \times \mathbb{R}^n) \quad \wedge \quad \alpha_{gcm}(M, F) \qquad (6a)$$

where α_{gcm} says that the $\varphi \in F$ are bijections from M onto $\mathbb{R}^n$,
that $\psi \varphi^{-1} \in G$ for $\varphi, \psi \in F$, and that $g \cdot \varphi \in G$ for $\varphi \in F$ and $g \in G$. These
requirements make F into a complete set of global coordinate

systems on the space M with respect to the group G of coordinate transformations. It can easily be proved that α_{gcm} is canonically invariant with respect to the typification in (6a).

The most important examples of gcm(n,G) applied in physics are those in which M is interpreted as spacetime (hence n = 4) and F is a class of preferred coordinate systems in spacetime with one of the following groups of coordinate transformations (cf. Ehlers 1973):

$$G_{new} \subset \begin{matrix} G_{gal} \subset G_{ros} \subset G_{kin} \\ \subset G_{aff} \\ G_{poi} \end{matrix} \subset G_{diff} \subset G_{top} \quad (7)$$

This hierarchy provides us with so many gcm(4,G...) as G... is one of the following transformation groups of $\mathbb{R}^4$: the direct product of the euclidean groups of space and time (G_{new}), the Galileo (G_{gal}) and Poincaré (G_{poi})group, the affine group (G_{aff}), the enlargements of G_{gal} allowing for accelerative translational (G_{ros},cf. Rosen 1972) and completely arbitrary (G_{kin}) rigid motions, the group of diffeomorphisms (G_{diff}) and of homeomorphisms (G_{top}) of $\mathbb{R}^4$. Besides the direct spacetime applications of the species of gcm other applications of interest are the (global) configuration spaces and their cotangent bundles (phase spaces) in classical mechanics. All these cases, however, and in fact the species of gcm in general have still to be enriched with further structures that can be used for a more complete description of physical reality. This is achieved in a second step by defining a species of global manifolds (gm) to be any extension of a species of gcm. According to the definition of extensions in section I this means that (6a) is supplemented by a formula

$$s \in \sigma(M) \wedge \alpha(M; F, s) \quad (6b)$$

such that their conjunction is a species of structures. Depending, as it does, on the parameters n and G, belonging to gcm(n,G), as well as on σ and α in (6b) a species of gm will be designated by gm(n,G,σ,α). With the introduction of the new structure s typified and axiomatized in (6b) all sorts of extensions of the previously mentioned physical applications of a gcm come within view. Outstanding examples are to be found in newtonian particle mechanics (based on G_{gal}), nonrelativistic quantum mechanics (usually based on G_{new},

but generalizable to G_{gal} and even G_{ros} , cf. Rosen 1972),
classical electrodynamics (based on G_{pol}), relativistic
quantum mechanics of the Dirac equation (based on G_{poi}) and
Einstein's gravitational theory (based on G_{diff}), the latter
being restricted to the global case. In all these cases the
structure s in (6b) would stand for the physical objects
that are under investigation, e. g. orbits of particles,
spacetime extensions of fields, probability functions for
quantities etc., and α of (6b) would, among other things,
state the essential physical laws that the objects have to
obey, e. g. a law of motion, a law of propagation etc.

For the moment I do not want to go into any details
about the question how the theories just mentioned actually
can be reconstructed as so many species of global manifolds.
The answer to this question is intimately connected with the
main problem to be treated in this section. If, for instance,
we were asked to give a formulation of Newton's law of gravitation for a system of mass points as a special case for (6b)
then a very common answer would be to write down the well
known gravitational equations in galilean coordinates, justifying this procedure by the remark that the equations are
invariant under any change of this kind of (galilean) coordinate systems. It is here where we meet the 'common invariance' of physical laws as invariance of certain 'concrete'
mathematical equations representing the laws in question
under an appropriate representation of a 'concrete' mathematical group. At the same time, the formulation of Newton's
law mentioned before would <u>not</u> be a formulation in terms of
species of gm <u>although</u> we feel that it may be equivalent to
such a formulation and that the reason for this is just the
'common invariance' of the equations representative for the
law.

We may therefore try to solve our main problem by seek-
ing a <u>characterization</u> of the species gm(n,G,σ,α) by means
of appropriate equivalents realizing the idea of a coordina-
te formulation of (6b). One thing, however, that must be
clearly understood about such an attempt is that it has no
unique, intrinsic solution but rather depends on a decision
as to how an object s $\in$ $\sigma(M)$ is going to be represented in
a coordinate system and, accordingly, how the group G of
coordinate transformations is going to be represented on
the set of representatives of the s. In the first characterization to be given we decide to take canonical representations throughout. Then, given n, G and σ, a canonical coordinate formulation ccf(n,G,σ,α') is a formula

$$s \in \sigma(R^4) \wedge \alpha'(s) \tag{8a}$$

where α' is <u>invariant</u> in the sense that

$$g \in G \wedge s' = g^\sigma \cdot s \wedge s \in \sigma(R^n) \dashrightarrow \alpha'(s) \leftrightarrow \alpha'(s') \tag{8b}$$

is provable. (8b) with g^σ as the canonical representation of g is here offerred as a condition of invariance in the usual sense, and (8a) will have to replace (6b) in the wanted characterization.

This characterization can most conveniently be exressed in terms of a relation between two formulas α and α'. Given n, G and σ the relation is to hold if and only if
(C_1) α extends the data to a species gm(n,G, ,), or — equivalently —
(4) holds for α with σ_0 and α_0 being given by (6a),
(C_2) α' extends the data to a canonical coordinate formulation, or — equivalently —   satisfies (8b),
(C_3) the formula

$$\varphi \in F \wedge s' = p^\sigma \cdot s \wedge s \in \sigma(M) \dashrightarrow \alpha(M; F, s) \leftrightarrow \alpha'(s') \tag{9a}$$

or — equivalently and slightly more to the point —

$$\varphi \in F \wedge s' = p^\sigma \cdot s \dashrightarrow s \in \sigma(M) \wedge \alpha(M; F, s) \leftrightarrow s' \in \sigma(R^4) \wedge \alpha'(s') \tag{9b}$$

is provable from (6a).
It turns out that this relation is itself invariant under two equivalence relations for α and α'. For α it is given by the provability of

$$s \in \sigma(M) \dashrightarrow \alpha(M; F, s) \leftrightarrow \alpha_1(M; F, s) \tag{10a}$$

from (6a), and for α' by the provability of

$$s \in R^n \dashrightarrow \alpha'(s) \leftrightarrow \alpha_1'(s) \tag{10b}$$

(from ZFT alone). The characterization will be possible only <u>modulo</u> these equivalence relations. It says (equivalence theorem 1) that the relation defined by (c) is one-to-one <u>modulo</u> (10) and that to every α satisfying (C_1) there exists $\overline{\alpha'}$ such that the relation holds and, conversely, that to every α' satisfying (C_2) an α exist such that the relation holds. An explicit solution of (9) for α, given α', is

$$\bigvee s'_1 \varphi : \varphi \in F \wedge s' = \varphi^6 \cdot s \wedge s' \in \sigma(R^n) \wedge \alpha'(s') \qquad (11a)$$

and for α', given α,

$$\bigvee M, F, s, \varphi : \varphi \in F \wedge s' = \varphi^6 \cdot s \wedge s \in \sigma(M) \wedge \alpha(M; F, s) \qquad (11b)$$

In the present case the right side of (11b) even reduces to $\alpha(R^n; G, s')$ and has been given the complicated formulation only for reasons that will become clear in the sequel.

The proof of equivalence theorem 1 is, although a bit tedious, a straightforward matter and need not be given here. The theorem clearly shows in what sense our main problem has been solved (under the present restrictions): On account of the one-to-one correspondence established by the theorem the common invariances (8b) of coordinate formulations are <u>exactly matched</u> by the canonical invariances (4) in species of structures. At the same time the theorem provides a <u>standard method</u> to produce reformulations of the mathematical part of a physical theory as a species of structures if a canonical coordinate formulation of that theory is known. This is the case, for instance, with many subtheories of classical mechanics, Newton's theory of gravitating mass points being the most prominent example. On the other hand, the usual coordinate formulation of electrodynamics, for instance, is not canonical. The group of coordinate transformations (G_{pot}) gets a tensor representation instead. To include such important cases in our reduction programme the equivalence theorem will have to be generalized.

Before this will be done one other remark should be made now. Equivalence theorem 1 is, of course, nothing but a methodologically purified version of the transition from the old way of coordinate geometry to the modern, abstract and coordinate-free approach to geometry. Besides showing the ordinary invariances becoming canonical invariances in species of structures in this transition the theorem also shows in what sense we get rid of coordinates. A coordinate formulation 1) of necessity picks out <u>one</u> of the preferred coordinate systems, but 2) leaves it completely arbitrary <u>which</u> one is chosen. As can be most clearly seen in (9b) what we get rid of is the arbitrariness of saying what has to be said in terms of one coordinate system. What we get <u>not</u> rid of is the very set of preferred coordinate systems out of which the choise has to be made. Far from becoming eliminated this set rather is introduced explicitly as a structure in a coordinate manifold. This is sometimes blurred by an otherwise

very convenient piecewise application of the standard method.
In most physical theories the structure s as well as what is
said about it in (6b) both are fairly complicated, and one
may wish to break them up in order to get simpler units. This
method consists of three parts: 1) a decomposition of (8b)
2) the transition of the pieces to (6b) by means of the equivalence theorem, and 3) their recomposition in accordance
with their original connection. Since as a rule no reference
to the preferred set F of coordinate systems is needed in the
third step this may give the impression that F, too, has been·
eliminated. This, however, is not the case since the second
step heavily depends on F.

The result of equivalence theorem 1 may not come as a
surprise for the simple reason that (8b) is a special case
of canonical invariance. Once we had decided to use the
canonical representation of G in the coordinate formulations
it was very suggestive to found the relation (c) on the canonical representation of F. But not all representations are
canonical. In fact the most important representations used in
physics (besides the canonical), namely the tensor and spinor
representations of the classical groups, are not. The question
therefore arises whether the coordinate formulations of physical theories based on these representations also have species of structures as equivalents and whether the invariances
belonging to the former again have the canonical invariances
in species of structures as their counterparts.

To answer this question for almost completely general
representations in coordinate formulations we introduce a
new class of extensions of species gcm(n,G) depending on three
additional external parameters. They are two scale terms σ
and σ' (each with one principal argument) and an arbitrary
representation r of G on $\sigma'(R^n)$ in the usual sense. (Like
G itself r is meant to be a term, available in ZFT, for which
it can be proved that it has the property just required.) A
species of <u>representations in a gcm</u> designated by rp(n,G,σ,
σ', r) – is then defined to be an extension of gcm(n,G)
with

$$\rho \in Pot(Pot(M \times R^n) \times Pot(\sigma(M) \times \sigma'(R^n))) \wedge \alpha_{rp}(M; F, \rho) \quad (12)$$

as the new typification and axiom. The axiom α_{rp} says that
ρ is actually a mapping

$$\rho : F \rightarrowtail bij(S_\rho; \sigma'(R^n)) \qquad (13a)$$

from F into the set of bijections from a set $\mathcal{S}_\rho \subseteq \sigma(M)$ (uniquely determined by ρ) onto $\sigma'(R^n)$ compatible with the representation r in the sense that

$$\varsigma(\psi) = r(\psi\varphi^{-1})\,\varrho(\varphi) \qquad (13b)$$

and with the canonical representation on $\sigma(M)$ in the sense that

$$\varsigma(\psi) = \varrho(\varphi)\,(\varphi^{-1}\psi)^\sigma \mid \mathcal{S}_\rho \qquad (13c)$$

for any $\varphi, \psi \in F$. Thus $\rho(\varphi)$ does for the objects $s \in \mathcal{S}_\rho$ what a coordinate system does for the points of M: As φ represents the latter in R^n so $\varrho(\varphi)$ represents the former in $\sigma'(R^n)$.

Species of representations are applied in physics in a peculiar way: A representation being an extension of a gcm just as, for instance, some physical field nevertheless does not have the latter's contingency. Rather there are some distinguished representations singled out on every gcm by means of a deduction. A deduction (here restricted to the case where base sets are kept fixed, cf. Bourbaki 1968, Ch. IV) relates two species of structures $(\sigma; \alpha)$ and $(\tau; \beta)$ by a term P that deduces the latter from the former in the sense that

$$P(X;s) \in \tau(X) \;\wedge\; \beta(X; P(X;s)) \qquad (14a)$$

together with the invariance condition

$$iso_{\sigma \times \tau}(X, s, t\,; X', s', t'; f) \;\rightarrow\; t = P(X;s) \leftrightarrow t' = P(X';s') \quad (14b)$$

is provable from $(\sigma; \alpha)$. The following is a deduction of $rp(n, G, \sigma, \sigma', r_{can})$ from gcm(n,G): $D_{can}(M;F)$ assigns to every $\varphi \in F$ its canonical image $\varphi^\sigma: \sigma(M) \rightarrowtail \sigma(R^n)$ defined by (2). This is the representation on which equivalence theorem 1 was founded. A second kind of distinguished representations deduced from gcm(n,G) presupposes that $G \subseteq G_{diff}$. They concern the tensors and tensor fields of any valence. Whereas in the case of the canonical representations we had $\mathcal{S}_{D_{can}} = \sigma(M)$ and $\sigma' = \sigma$, in the tensor representations $\mathcal{S}_{D_{ten}}$ is restricted to the bundle of tensors of a given valence (k,l) and $\sigma'(R^n) = R^n \times R^{n^{k+l}}$. For tensor fields the corresponding power sets have to be used. The representations themselves are defined in the usual way by

giving the tensors and tensor fields their components in a
given coordinate system. A third kind of distinguished representations concerns affine connections. As in the tensor
case the representation is confined to a special kind of
objects, and it is $\sigma'(\mathbb{R}^n) = \text{Pot}(\mathbb{R}^n \times \mathbb{R}^{n^3})$.

Starting now from species gcm(n,G) and rp(n,G, σ, σ' ,r)
and a deduction D of the latter from the former we can generalize equivalence theorem 1 in the following way: Replace
in the formulas (8) - (11)

$$
\begin{array}{lll}
\sigma'(\mathbb{R}^n) & \text{for} & \mathcal{V}(\mathbb{R}^4) \\
\mathcal{V}(q) & \text{for} & q^\sigma \\
S_{\partial(M;F)} & \text{for} & \sigma(M) \\
D(M;F)\cdot\mathcal{V} & \text{for} & q^\sigma
\end{array}
\qquad (15)
$$

Then the equivalence theorem holds <u>ceteris paribus</u> with respect to the new entities, i. e. given n, G, σ, σ', r and
D as above the relation (c) establishes a one-to-one correspondence <u>modulo</u> (10) between all species gm(n,G, σ, α)
and all coordinate formulations cf(n,G, σ', α') (equivalence
the orem 2). The essential generalization that has been obtained is the step from canonical to (almost) arbitrary representations and corresponding invariances on the side of
the coordinate formulations <u>retaining</u> thereby canonical invariance on the side of the species of structures.

3. COVARIANCE

Up to this point the considerations concerning the reduction
of the ordinary invariances occurring in coordinate formulations of physical theories to canonical invariance as part
of the definition of species of structures were confined to
the topologically trivial, global case where the space M is
homeomorphic to $\mathbb{R}^n$. In order to gain the full generality
in which the theory of manifolds is used to be developed in
mathematics and even applied in physics we would now have
to take a final step by introducing the <u>local viewpoint</u>. As
will be seen in a moment this causes no problems as far as
the concept of a manifold based on a coordinate manifold is
concerned. However, as soon as one proceeds from this new
generalized basis in order to get a corresponding generalization of the equivalence theorem considerable difficulties
come up. Already the further conceptual apparatus for a new
formulation of the theorem would demand an amount of technical detail that would be beyond the scope of the present paper.

I will therefore leave the matter at the stage that has been
achieved in the previous section and now switch over to the
concept of covariance.

The species of global coordinate manifolds gcm(n,G) can
be generalized by replacing the requirement that G be a group
of global homeomorphisms of R^n by the weaker assumption that
G is only a pseudogroup of local homeomorphisms in R^n (cf.
Iyanaga and Kawada 1977, 92 D). The idea of a pseudogroup of
homeomorphisms generalizes that of a group of homeomorphisms
of R^n by allowing variable domains and ranges. The local
viewpoint that is opened in this way sometimes is emphasized
by the requirement that every restriction of an element of
G to an arbitrary open subset of its domain also belongs to
G (cf. Iyanaga and Kawada 1977, 108 Z; Kobayashi and Nomizu
1963, p. 1). Since this excludes groups of transformations
as special cases of pseudogroups I adopt the more general
concept mentioned first. The global case considered so far
is recovered by the requirement that all transformations
have a common domain (mostly R^n itself). It occurs if and
only if the pseudogroup is a group. The groups G... in (7)
all have obvious counterparts G... meeting the requirement
of strict locality mentioned above.

Given n and G as a pseudogroup the species of <u>coordinate
manifolds</u> cm(n,G) is then defined in formal analogy to (6a).
But the new axiom α_{cm} would only say that F is a maximal set
of local coordinate systems compatible with the transformations of G. The species of gcm considered so far are those
species of cm where the transformations of G have a common
domain and hence the spaces M are homeomorphic to (an open
subset of) R^n . Again on the general level a species of
<u>manifolds</u> mf(n,G, σ , α) can then be introduced to be any
extension (6b) of a species cm(n,G). This concept is sufficiently general to include all kinds of manifolds that are
known from differential geometry such as riemannian manifolds,
manifolds with an affine connection and those that carry
additional fields of all sorts that have been studied in
theoretical physics.

Being thus prepared to approach covariance on the most
general differential geometric level there is even reason to
step on the more abstract level of arbitrary species of structures and develop our concept on it. As was already announced
in the introduction our access to covariance will be guided
by the idea of equivalent mathematical formulations of one
and the same physical theory. Since species of structures
lend themselves to such formulations we have to ask for an

adequate concept of equivalence for them. This is readily at
hand: Given two species $(\sigma;\alpha)$ and $(\sigma_1;\alpha_1)$ as well as
deductions P and P_1 of $(\sigma_1;\alpha_1)$ from $(\sigma;\alpha)$ and <u>vice versa</u>
such that

$$P(X; P_1(X;s_1)) = s_1$$
$$P_1(X; P(X;s)) = s \tag{16}$$

are provable from $(\sigma;\alpha)$ and $(\sigma_1;\alpha_1)$ respectively $(\sigma;\alpha)$
and $(\sigma_1;\alpha_1)$ are called <u>equivalent</u> with respect to P and
P_1. That of two species of structures equivalent in this sense
either both or neither of them can be usedto formulate a physical theory hinges on questions of interpretation that can
not discussed in this paper (cf. Ludwig 1978). It is, however,
safe to say that, given $(\sigma;\alpha)$ together with a physical interpretation, the move to an equivalent $(\sigma_1;\alpha_1)$ by means of
equivalence terms P and P_1 can not affect the content of the
theory <u>if</u> the interpretation is retained, and it seems plausible that by using P and P_1 the interpretation can always be
transferred from $(\sigma;\alpha)$ to $(\sigma_1;\alpha_1)$ to yield an interpretation of $(\sigma_1;\alpha_1)$ in the same sense as the original interpretation of $(\sigma;\alpha)$. In Ludwig 1978 such a transfer is discussed
even for the weaker case where the base sets can be different
and P_1 is replaced by the assumption that P is conservative
in the sense that every structure $\langle X;t\rangle$ of species $(\tau;\beta)$
is isomorphic to a structure $\langle X;P(X;s)\rangle$ with an $\langle X;s\rangle$ of species $(\sigma;\alpha)$. For treatment of problems of covariance, how-
ever, the stronger and simpler concept of equivalence seems
to be sufficient.

Let now Σ and Σ_1 be two species of structures and P a
deduction of Σ_1 from Σ. Furthermore, let Σ^* and Σ_1^* be two extensions of Σ and Σ_1 respectively and equivalent with respect
to the equivalence transformation

$$F_1 = P(M; F), \qquad F = P_1(M; F_1, s_1)$$
$$s_1 = Q(M; F, s), \qquad s = Q_1(M; F_1, s_1) \tag{17}$$

where the <u>given</u> deduction P appears in the left upper corner.
Here the notation is already adapted to species of mf but the
actual assumptions are meant to be quite general. The following scheme with P^* for (17) may help to apprehend the given
data and their mutual relations:

$$\begin{array}{ccc}
\Sigma & \xrightarrow{\quad P \quad} & \Sigma_1 \\
\downarrow & deduction & \downarrow \; extension \\
& equivalence & \\
\Sigma^* & \xleftarrow{\quad P^* \quad} & \Sigma_1^*
\end{array} \qquad (18)$$

We now call Σ_1^* a <u>covariant version</u> of Σ^* and – conversely –
Σ^* a <u>reduced version</u> of Σ_1^*, both with respect to the remaining data Σ, Σ_1, P and P^* . It should be borne in mind
that the concepts of covariance and reduction thus defined
strictly speaking relate six syntactical entities to each
other. In actual practise, however, we can think of the three
data in the upper row of (18) as being 'kept fixed'.

This will become clear if we now specialize the general
setting in (18) by assuming Σ and Σ_1 to be two species
cm(n,G) and cm(n,G_1) with $G \subseteq G_1$. We then have the following
natural deduction P of cm(n,G_1) from cm(n,G): Given a
cm $\langle M;F \rangle$ of the species cm(n,G) the set F_1 = P(M;F) is the
iniquely determined complete set of coordinate systems on M
with respect to G_1 satisfying $F \subseteq F_1$. More specificly, since
the days when the general theory of relativity was born the
cases of prevailing interest became those in which G was one
of the classical groups and G_1 the pseudogroup G_{diff}° . And
since then it became a prevailing tendency in theoretical
physics to show of as many spacetime founded physical theories as possible that they could be reformulated as certain
extensions of the species of differentiable manifolds.

Before extending the exemplification in this direction
also to the lower row of (18) two problems shall be formulated that pose themselves in the situation when only the upper
row of (18) is fixed. To be as precise as possible their
wordings will again be given in general terms. They are: (co)
If in addition to Σ , Σ_1 and P of (18) an extension Σ^* of Σ is
given can we then always find Σ_1^* and P^* such that Σ_1^* is a
covariant version of Σ.(re) If in addition to Σ, Σ_1 and P of
(18) an extension Σ_1^* of Σ_1 is given can we then always find
Σ^* and P^* such that Σ^* is a reduced version of Σ . The two
problems will be called the <u>problem of covariance</u> and of
<u>reduction</u> respectively. There is obviously a certain duality
between them. However, on account of the asymmetry of (18)
in its upper row the two problems receive entirely different
answers.

The answer to the problem of covariance is positive and
trivial. Assuming that the axiomatics of the given species
Σ, Σ_1 and Σ^* (typification and axiom proper) are

$$\Sigma : \bar{\alpha} (M;F), \qquad \Sigma_1 : \bar{\alpha}_1 (M;F_1)$$

$$\Sigma^+ : \bar{\alpha} (M;F) \wedge \bar{\beta}(M;F,s) \tag{19a}$$

then with $P(M;F)$ as the given deduction of Σ_1 from Σ the species Σ_1^* that we are looking for can be chosen to be

$$\Sigma_1^* : \bar{\alpha}_1(M;F_1) \wedge \bar{\alpha} (M;F) \wedge \bar{\beta}(M;F,s) \wedge F_1 = P(M;F) \tag{19b}$$

where the variables have already been chosen such that the equations of (17) become identities except, of course, the one that is prescribed by P. Let us exemplify this solution in the context of species of manifolds taking up the settings from the preceding paragraph. Suppose then that we want to have a covariant version of ordinary electrodynamics. In this case $\bar{\alpha}$ would axiomatize Minkowski spacetime, $\bar{\alpha}_1$ the species of differentiable manifolds, and $\bar{\beta}$ (as part of Σ^*) would be a reformulation of Maxwell's equations in terms ofspecies of structures according to equivalence theorem 2. The trick of getting at a covariant version of Σ^* now consists in adding s (the Maxwell field) <u>as well as</u> a set F of distinguished coordinate systems (the Lorentz frames) to the set F_1 of arbitrary coordinate systems in Σ_1. Precisely this is expressed by $\bar{\alpha} \wedge \bar{\beta}$ and $F_1 = P(M;F)$ which in the present case turns out to be equivalent to $F \subseteq F_1$.

In view of solution (19b) in all its triviality there may be two kinds of objections against our version (co) of the problem of covariance. It may,firstly, be objected that the nontrivial problem will be an analogue of (co) which exclusively deals with <u>coordinate formulations</u>. It may, for instance, seem a nontrivial problem to find a coordinate formulation of the ordinary Maxwell equations that is invariant under G_{diff} . However, given the results of section II the objection can easily be met by combining those results with (19b): Equivalence theorem 2 does provide for a coordinate formulation equivalent to (19b). In the Maxwell case this coordinate formulation turns out to say about $\langle F', s' \rangle$ that F' is a right coset of G_{poi} in G_{diff} and s a solution of the equations that result from the Maxwell equations by transforming them with any $g \in F$, - the former equations being uniquely determined by F on account of the invariance of Maxwell's equations under G_{poi} .

According to the second objection it would be pointed out that (co) could be made nontrivial if only <u>additional requirements</u> were imposed on the species Σ_1^* that is to be

found. In the differential geometric context laid down above
it could, for instance, be required that the new structure
s_1 of $mf(n,G_1)$ extending the coordinate manifold $\langle M;F \rangle$ of
$cm(n,G_1)$ has to be a cartesian product of 1) curves in M,2)
tensor fields on M, 3) one affine connection on M, and nothing
else. This would be a perfectly clear requirement, and the
trivial solution (19b) would certainly not satisfy it. Moreover, the new problem (co) would be nontrivial in the twofold
sense that it would not any more have a positive answer for
every given Σ^{*}, and if it has, the answer may very well be
nontrivial, - at least not as trivial as (19b).
　　To this objection I wholeheartedly agree. At the same
time, the existence of the unrestricted formulation (co) will
now be justified as the necessary background for it. It is
hard to find out whether there is such a thing as the prevailing opinion about the concept of covariance and problems
connected with it. But I had always the impression that if
there is then part of it is the tendency to view covariance
as something which can be had - and in a sense even trivially
had - at all events. On the other hand, I have never seen any
proof or even attempt to prove that this view is correct. In
order to explore under what circumstances a proof in the
strict sense would be possible the version (co) and its solution (19b) were produced. And it may be repeated with regard to the first objection that together with the results
of section II both the problem (co) and its solution cover
corresponding questions concerning invariance as they are
usually mixed up with covariance.
　　Now, adherents of the triviality thesis about covariance
becoming aware of the solution (19b) might still not feel
themselves confirmed in their intuitions. Regarding the
additional requirement objection they would perhaps point out
that, once differential geometry had become of principal
physical interest as a consequence of the empirical success
of Einstein's gravitational theory, covariant versions meeting
the objection have been found for the formalisms of all classical geometrical and physical theories. And since these were
theories of widely differing contents the property of covariance shared by all of them can not be of much physical
importance. In other words, the triviality of covariance
consists in the lack of any contribution to the content of a
physical theory. Before commenting on this argument something has to be said about the fact that it alludes to.
There could indeed be made a long list of embeddings of geometrical and physical theories in the general differential

geometric context mentioned in the additional requirement objection (or something similar to it). But in order to discuss
the consequences of this fact with sufficient rigour the
preliminary question had to be answered precisely in what
sense covariance had been established in all these cases. It
seems that the concept of covariance proposed above will cover
all the relevant cases known from the literature (see especially the more recent work done in Havas 1964, Trautman
1967, Künzle 1972, Misner et al. 1973, Ch. 12). In all these
cases species $gm(n, G^{b}_{diff})$ submitted to the restrictions of
our second objection are produced as equivalents of species
$gm(n, G)$ with $G \subseteq G^{b}_{diff}$ that function as the 'original' formulation of a physical theory. The need for a separate concept
of covariance, different from that of invariance, becomes
particularly clear in a study of the examples: Since the invariance properties of the given species $gm(n, G)$ as a rule
are already exhausted by G the covariant version $gm(n, G^{b}_{diff})$
can not be produced without drastic changes that even touch
upon the type of the structures involved.

As to the argument that covariance is to weak a property
to make any considerable contribution to the content of a
physical theory it has to be admitted that even under the
restrictions of the additional requirement objection the class
of species $gm(n, G)$ with covariant versions in G^{o}_{diff} is very
comprehensive. But it had to be pointed out that in the light
of the utterly trivial solution (19b) the class in question
is already restricted, the answer to (co) is not any more
positive throughout, and therefore the historical fact described in the last paragraph is not trivial in the sense
that it could not have been foreseen a priori. Moreover,
particularly with Einstein's postulate of general covariance
in view one should look at the whole matter also in the
opposite direction that is provided by the problem of reduction (re). As has been already remarked, contrary to the
problem (co) the problem (re) does not have a positive answer
for every case of given data. Thus in the differential geometric context, given $gm(n, G_4)$ (for Σ_1^{*}) and $cm(n, G)$ with
$G \subseteq G_4$ (for Σ) it may not be possible to find an extension
$gm(n, G)$ equivalent to $\overline{gm(n, G_4)}$. The latter species would
then be irreducible in the sense that it is irreducible to
G for every G included in but different from G_4. With this
concept at hand the postulate of 'general covariance' can
easily be made nonvacuous by reinterpreting it as the want
for species $gm(n, G^{o}_{diff})$ that are absolutely irreducible.
Questions of real frames of reference putting aside this

interpretation comes fairly close to what Einstein had in mind by looking for physical laws in 'arbitrary' coordinate systems: The laws were meant to be such that they would not prefer any subset of coordinate systems. Now, in a well defined sense this is the case for absolutely irreducible species $gm(n, G^0_{dif})$, and the impressive thing about Einstein's gravitational theory was that it was the first physical theory using such species. With respect to irreducibility 'arbitrary' coordinates really took over exactly the same role in the new gravitational theory (and its general extensions) as had the Lorentz frames (viewed merely as coordinate systems) in physical theories based on special relativity.

Erhard Scheibe
Philosophisches Seminar
Georg-August-Universität Göttingen

[1] The completion of this paper was made possible by a Visiting Fellowship at the Center for Philosophy of Science at the University of Pittsburgh. The author wants to express his gratitude to chairman and director of the Center, Professor Grünbaum and Professor Laudan, for the kind invitation and generous hospitality.

REFERENCES

Bourbaki, N.: 1968, Elements of Mathematics. Theory of Sets, Paris.

Ehlers, J.: 1973, The Nature and Structure of Spacetime, in The Physicist's Conception of Nature, Ed. J. Mehra, Dordrecht, 71-91.

Havas, P.: 1964, 'Four-Dimensional Formulations of Newtonian Mechanics and their Relation to the Special and the General Theory of Relativity', Review of Modern Physics 36, 938-65.

Iyanaga, S., and Y. Kawada (eds.): 1977, Encyclopedic Dictionary of Mathematics, Cambridge, Mass.

Kobayashi, S., and K. Nomizu: 1963, Foundations of Differential Geometry, vol. I., New York.

Künzle, H.P.: 1972, 'Galilei and Lorentz Structures on Space-Time: Comparison of the corresponding Geometry and Physics'. Annales de l' Institut Henri Poincaré XVII, 337-62.

Ludwig, G.: 1978, The Basic Structures of a Physical Theory (in German), Berlin.

Misner, Ch. W., Thorne, K.S., and J.A. Wheeler: 1973, Gravitation, San Francisco.

Rosen, G.: 1972, 'Galilean Invariance and the General Covariance of Nonrelativistic Laws'. American Journal of Physics 40, 683-7.

Scheibe, E.: 1978, 'On the Structure of Physical Theories', in The Logic and Epistemology of Scientific Change, Eds. I. Niiniluoto and R. Tuomela. Acta Philosophica Fennica XXX, Iss. 2-4, 205-24.

Shoenfield, J.F.: 1967, Mathematical Logic, Reading, Mass.

Stegmüller, W.: 1979, The Structuralist View of Theories, Berlin.

Trautman, A.: 1967, 'Comparison of Newtonian and Relativistic Theories of Space-Time', in Perspectives in Geometry and Relativity, Ed. B. Hoffmann, Bloomington, Ind., 413-25.

Trautman, A.: 1972, 'Invariance of Lagrangian Systems', in General Relativity, Ed. L.O. Raifeartaigh, Oxford, 85-99.

W.Balzer[0]

THE ORIGIN AND ROLE OF INVARIANCE IN CLASSICAL KINEMATICS

By classical kinematics (CK) I understand the theory the models of which have the form $\langle P, \mathbb{R}, s \rangle$, where P is a non-empty set (of points or 'parti= cles'), $\mathbb{R}$ is the set of real numbers,and

$s:P \times \mathbb{R} \rightarrow \mathbb{R}^3$ is smooth in its second argument. s is called position function and $\mathbb{R}$ denotes time, so $'s(p,t)= \langle \alpha_1, \ldots, \alpha_3 \rangle'$ has to be read as 'point p at time t is in position $\langle \alpha_1, \alpha_2, \alpha_3 \rangle'$.[1]

It is well known that models of classical kinematics are invariant under spatial- and time-displacements,spatial rotations,and combinations of these three kinds of transformations.More pre= cisely,if $\langle P, \mathbb{R}, s \rangle$ is a model and if

$s':P \times \mathbb{R} \rightarrow \mathbb{R}^3$ is defined by

there are a real,orthogonal 3 × 3-matrix
$\alpha, f \in \mathbb{R}^3$ and $b \in \mathbb{R}$ such that for all   (1)
$t \in \mathbb{R}$ and $p \in P$: $s'(p,t)= \alpha s(p,t+b)+f$,

then $\langle P, \mathbb{R}, s' \rangle$ again is a model of classical kine= matics.Let me call transformations of this kind 'space-time transformations'.What has been just said then can be expressed by saying that CK is invariant under space-time transformations.

A philosopher of science would be satisfied with stating that space-time transformations are just those transformations under which CK's axioms are invariant.But this is not the case.CK in fact is invariant under a much bigger class of transformations.How then does it come that physicists have concentrated just on space-time transformations? I try to answer this question by investigating the origin and role of space-time transformations.It turns out that philosophers of science can be satisfied by showing CK to be a

149

D. Mayr and G. Süssmann (eds.), Space, Time, and Mechanics, 149–169.
Copyright © 1983 by D. Reidel Publishing Company.

theoretization of underlying theories of space and
time.If those underlying theories are taken into
account when formulating CK's axioms all but space-
time transformations are excluded.[2]

    Intuitively,I can summarize my results on the
origin and role of invariance under space-time
transformations in CK in form of a thesis.

    The invariance of CK under space-time trans=
    formations originates from concrete obser=
    vations that distances in space and time remain
    the same under spatial- and time-displacements
    as well as under spatial rotations of 'space-
    time'-coordinate systems.The actual role of
    this invariance consists of two components.
    First,invariance gives us a certain freedom of
    choice for coordinate systems.Second,very often
    one can execute 'active' counterparts of trans=
    formations of coordinate systems -i.e.active
    transformations of the physical system itself
    relative to a fixed coordinate system- with
    the effect of not changing the system.Thus in=
    variance serves as a guide in the application
    of active space-time transformations.

I will first tell a science fiction story about
how theories of space and time might have de=
veloped in order to optimally please philosophers
of science thinking about space-time invariance.
Only after this story I will try to see how it
increases the probability of my thesis.

1. A SCIENCE FICTION STORY

In some culture on our planet the art of dealing
with distances and motions might have developed
as follows.By using a certain kind of measuring
instruments,say 'rigid rods',in order to compare
distances people recognize that what they are
doing can be described by 'bringing ends of (or
marks on) rigid rods to coincide with marks or
particles or marks on bodies','testing whether
rods are straight' and by 'testing whether rods
are equally long'.People find out that certain
statements about these operations regularly turn
out as true.They introduce some abbreviations by
referring to a set P of points (or particles or

marks),a relation $\underline{b}$ ('$\underline{b}abc$' meaning that point b
on a rod is situated between points a and b),and
to a relation $\equiv$ ('$ab\equiv a_1 b_1$' meaning that the rod
with end points a and b is just as long as the
rod with end points a_1 and b_1).The statements
which turn out to be true are just those state=
ments which in Tarski (1959) are taken as axioms
for a system of Euclidean geometry.[3] In short,
people find out that their handling spatial dis=
tances yields a realization (modulo idealization)
of some structure $\langle P,\underline{b},\equiv\rangle$ which is a model of
Euclidean geometry.

At the same time mathematics is developed and
people come to know real numbers.It turns out as
very effective to compare lengths of rigid rods
(or distances) by assigning to each rod a real
number such that the congruence relation under
this assignment is 'represented' by equality of
real numbers,and the betweenness relation is
'represented' by a kind of additivity of the
numbers assigned to the rods.[4] Finally,philosophers
of science come into being -wondering about and
questioning this development.They prove that,up
to the conventional choice of assigning the real
number 1 to some arbitrary rigid rod,such an
assignment of real numbers to rods is uniquely
determined by the statements about $\underline{b}$ and $\equiv$ and
the abovementioned representation properties.They
introduce a theory,the 'theory for metrization of
space' the models of which are entities of the
form $\langle P,\underline{b},\equiv,d\rangle$,where $\langle P,\underline{b},\equiv\rangle$ is a model of
Euclidean geometry,and $d:P^2\to\mathbb{R}$ is such that (1)
$\langle P,d\rangle$ is a metric space and (2) for all a,b,a',b'
$\in$ P: $\underline{b}aba'\leftrightarrow d(a,b)+d(b,a')=d(a,a')$ and $ab\equiv a'b'$
$\leftrightarrow d(a,b)=d(a',b')$.Here,d is the 'assignment of
numbers to rods' referred to above.d is called
distance function and the set of models $\langle P,\underline{b},\equiv,d\rangle$
of the theory for metrization of space is de=
noted by M1.

People further recognize that spatial de=
scriptions become much simpler if,instead of
stating all distances between all points involved,

they state for each point its distances to three
distinguished,eventually new,points which are the
same for the whole system.They introduce coordi=
nate systems (CS) for models in M1.A CS for x=
$\langle P,\underline{b},\equiv,d\rangle$ is just a tuple

$$y=\langle a_o,\ldots,a_3\rangle \in P^4$$

such that (1) for $0 \leq i \neq j \leq 3$: $a_i \neq a_j$,(2) for $1 \leq i \leq 3$
$d(a_o,a_i)=1$,and (3) for $1 \leq i \neq j \leq 3$:

$$d(a_i,a_o)^2+d(a_o,a_j)^2=d(a_i,a_j)^2.$$

a_o is called the origin and $a_1,\ldots,a_3$ are points
forming right angles with a_o (requirement (3)
above),and having distance 1 from a_o.Right angles
are expressed by Pythagoras' formula.The above=
mentioned simplification of spatial description
then is achieved by introducing in $x=\langle P,\underline{b},\equiv,d\rangle$
via a CS $y=\langle a_o,\ldots,a_3\rangle$ a unique function

$s:P \to \mathbb{R}^3$ such that (1) $s(a_o)=0$,(2) $s(a_i)=\mathcal{M}_i$ for
$1 \leq i \leq 3$,and (3) $\|s(a)-s(b)\|=d(a,b)$ for all $a,b \in P$.
Here $\mathcal{M}_i$ are the unit vectors and $\|\cdot\|$ is the
Euclidean norm on $\mathbb{R}^3$.s is unique because in a
model of Euclidean geometry some point $a \in P$ with
respect to y uniquely determines the four
distances $d(a,a_i)$ (i=0,...,3),and these,by con=
ditions (1)-(3),can be used to solve the equations
$\|s(a)-s(a_i)\|=d(a,a_i)$ for s(a).In this way,instead
of describing spatial relations by stating all
distances among all points,people have to give
only one vector s(a) for each point a in order to
obtain the same amount of information.And as the
number of particles increases,so does the amount
of simplification gained by this method.
     Now people make the following experience.If
they describe spatial relations of some concrete
object via a distinct CS y and afterwards via
another CS y' which does not move relative to y
then the resulting functions s and s' are connec=
ted by a space-transformation,i.e.there exists a

real,orthogonal 3 x 3-matrix α and a vector $\phi \in \mathbb{R}^3$
such that for all a $\in$ P: s'(a)=αs(a)+ϕ.Note
that this fact is observed,not deduced.

After these achievements there develops a
need for comparing processes or changes of various
kinds.By choosing a certain sort of measuring
instruments,for instance 'clocks',people are able
to compare different processes with each other by
referring to the initial and final events of those
processes.They recognize that what they are doing
can be described by 'bringing into coincidence
the initial and final events of processes','de=
ciding whether an event in a process preceeds an=
other one in that process','comparing different
processes' and 'concatenating two processes in
order to obtain a longer one'.They introduce some
abbreviations: a set T of initial events,final
events,and other clearly distinguishable events
occurring in processes,a relation $\lessgtr$ on such
events ('t$\lessgtr$t$_1$' meaning that event t preceeds
event t$_1$ in some process),a relation $\lesssim$ among
processes ('tt'$\lesssim$t$_1$t$_2$' meaning that the process
with initial and final events t and t' is not
longer than a process with initial and final
events t$_1$ and t$_2$),and an operation o among pro=
cesses ('tt'o t$_1$t$_2$=t$_3$t$_4$' meaning that the process
with initial and final events t$_3$ and t$_4$ is the
result of concatenating the processes with initial
and final events tt' and t$_1$t$_2$,respectively).By in=
vestigating various kinds of processes people
find out certain regularities.Using their abbre=
viations the statements expressing these regu=
larities are roughly the axioms for positive,
closed extensive systems (see Krantz et al.(1971),
p.73).As in the case of theories about space they
find it convenient to assign real numbers to pro=
cesses and to compare the latter by means of the
former.Again,this assignment has to 'represent'
the essential properties of the relations among
processes in a natural way by real numbers: $\lesssim$ is
represented by $\leq$ and o by +.And again,mathe=
maticians can prove that such assignments are
uniquely determined up to choice of a unit.In
short,a 'theory for metrization of time' is intro=

duced the models of which have the form
$\langle T, \leqslant, \preccurlyeq, \circ, \Im \rangle$,where (1) T is a set containing
at least two elements,(2) $\leqslant \subseteq T \times T$ is a linear
ordering which is dense,'continuous' and 'sepa=
rable',(3) $\preccurlyeq \subseteq T^4$ is such that for all t,t',t''
$\in$ T: $\langle t', t'' \rangle \preccurlyeq \langle t, t \rangle \rightarrow t'=t''$ and $\langle t, t \rangle \preccurlyeq \langle t', t'' \rangle$,
(4) $\circ : T^4 \rightarrow T^2$ is such that for all t,t',t'' $\in$ T:
$\langle t', t'' \rangle \circ \langle t, t \rangle = \langle t', t'' \rangle$ and $\langle t, t \rangle \circ \langle t', t'' \rangle =$
$\langle t', t'' \rangle$,(5) for D:=$T^2 \setminus \Delta$(T): $\langle D, \preccurlyeq_{/D}, \circ_{/D} \rangle$
is a positive,closed extensive structure in the
sense of Krantz et al.(1971),p.73,(6) $\Im$:T$\times$T$\rightarrow \mathbb{R}$
is such that $\langle T, \Im \rangle$ is a metric space and for all
$t, t', t_1, t_2, t_3, t_4 \in$ T: (6.1) $t \leqslant t' \leqslant t_1 \rightarrow \langle t, t' \rangle \circ$
$\langle t', t_1 \rangle = \langle t, t_1 \rangle$,(6.2) $\langle t, t' \rangle \preccurlyeq \langle t_1, t_2 \rangle \leftrightarrow \Im(t, t') \leqslant$
$\Im(t_1, t_2)$,and (6.3) $\langle t, t' \rangle \circ \langle t_1, t_2 \rangle = \langle t_3, t_4 \rangle \leftrightarrow$
$\Im(t, t') + \Im(t_1, t_2) = \Im(t_3, t_4)$.
Here Δ(T) is the diagonal in T and $f_{/D}$ denotes
f,restricted to D.Let M2 be the set of all such
models.
          As a matter of economy people introduce
coordinate systems which in this case are very
simple: a CS for $\langle T, \leqslant, \preccurlyeq, \circ, \Im \rangle$ consists of just
two events $\langle t^0, t^1 \rangle$ such that $t^0, t^1 \in$ T and $t^0 \leqslant$
t^1 (t^0 is something like 'the birth of christ'
and t^1 marks the end of the first unit of time
'after' t^0).With the help of a CS they can com=
pare the lengths of processes not by referring to
the 'distances' between initial and final events
but by referring to distances of events from t^0.
If $\langle t^0, t^1 \rangle$ is a CS for x=$\langle T, \leqslant, \preccurlyeq, \circ, \Im \rangle \in$ M2 then
there exists a unique function Θ :T$\rightarrow \mathbb{R}$ such that
(1) $\Theta(t^0)=0$,(2) $\Theta(t^1)=1$,(3) for all t,t'$\in$ T:
$t \leqslant t' \leftrightarrow \Theta(t) < \Theta(t')$,and (4) for all t$\in$ T: $|\Theta(t)|$
= $\Im(t, t^0)$. Θ here is the analogue to the s intro=
duced before in connection with space.That is,
Θ(t) gives the 'position' of event t with respect

to the CS $y=\langle t^0,t^1 \rangle$ for x.So by means of a CS
statements about comparison and lengths of pro=
cesses can be formulated in terms of events and
'coordinates' of events.

When the theory for metrization of time and
coordinate systems are introduced the following
is observed.If some concrete process and its de=
velopment in time is described from a CS
$\langle t^0,t^1 \rangle$ and also from a different CS $\langle \bar{t}^0,\bar{t}^1 \rangle$ then
the resulting functions θ and θ' are connected
by a time-transformation,i.e.there is some $\alpha \in \mathbb{R}$
such that for all $t \in T$: $\theta'(t)=\theta(t)+\alpha$.This is an
experimental result.

As soon as these two theories exist peoples'
interests turn to motions.Motions are certain
kinds of processes involving particles which in
themselves do not change.Such a process consists
of a relative change of different particles
constituting the process relative to each other.
With the help of spatial- and time-measurements
they can find out spatial distances between the
particles at various 'times'.On the other hand,
people also can measure these times and their
'distances' and start to talk about the rate of
spatial change with respect to a certain process
or the length of a process (or an 'interval of
time').But when they try to formulate this more
precisely in terms of the distance functions
already available they run into difficulties.If
they express spatial change by differences of
distances and periods of time by distances of
events how then they assure that the events are
precisely those at which the distances were mea=
sured? Since they have no term to express this in
their vocabulary and since they want to do so they
introduce a new term,a four-place function
$d:P^2 \times T^2 \to \mathbb{R}^2$ with the following meaning.
$'d(a,b,t,t')=\langle \alpha, \beta \rangle'$ means that the spatial
distance between a and b measured at event t is
α and the 'time-distance' between t and t'
measured at point a is β .Clearly,with this con=
cept they can formulate all kinds of statements
about motions and rates of spatial change in time.
Immediately they recognize that this new d 'con=

tains' the old d and $\mathcal{J}$ .For if the two time-argu= ments are fixed,say by t and t',one obtains a function $d_{tt'}:P^2 \to \mathbb{R}$,defined by $d_{tt'}(a,b)=$ $\pi_1(d(a,b,t,t'))$ (π_i is the projection on the i-th component).And by fixing the two spatial arguments,say by a and b,one obtaines a function $d_{ab}:T^2 \to \mathbb{R}$,defined by $d_{ab}(t,t')=\pi_2(d(a,b,t,t'))$. Also it is noticed that the non-metrical concepts of space,namely $\underline{b}$ and $\equiv$,must be adjusted in order to describe actual operations when time matters.Thus a new theory is introduced,the 'theory for metrization of space-time'.Its models are of the form

$$\langle T,P,\leqslant ,\preccurlyeq ,o ,\underline{b},\equiv ,d \rangle$$

where (1) T and P are sets,(2) $d:P^2 \times T^2 \to \mathbb{R}^2$, (3) for all a,b$\in$ P: $\langle T,\leqslant ,\preccurlyeq , o ,d_{ab}\rangle$ is in M2, (4) $\underline{b}\subseteq T \times P^3$ and $\equiv \subseteq T \times P^4$, (5) for all t,t',t'' $\in$ T: $d_{tt'}=d_{tt''}$ and $\langle P,\underline{b}_t , \equiv_t ,d_{tt'}\rangle$ is in M1 , (6) there are t,t'$\in$ T (t$\neq$t') such that for all a,b,c,e$\in$ P: $d_{ab}(t,t')=d_{ce}(t,t')$.Here,$\underline{b}_t$ and $\equiv_t$ are just $\{\langle a,b,c\rangle / \langle t,a,b,c\rangle \in \underline{b}\}$ and $\{\langle a,b,a',b'\rangle / \langle t,a,b,a',b'\rangle \in \equiv\}$ .In this theory the concepts T, $\leqslant ,\preccurlyeq$ and o concerning time are just as before.On the other hand,the spatial relations $\underline{b}$ and $\equiv$ have got an additional argument for time instants. From these time dependent spatial relations one obtains the original ones by considering 'cuts' $\langle P,\underline{b}_t , \equiv_t \rangle$ at certain instants t.Requirement

$d_{tt'}=d_{tt''}$ says that spatial relations as repre= sented by d depend only -if at all- on the first time-argument.The set of models of this theory is called M3.

In models of M3 space and time are connected as far as necessary by the introduction of a time-argument for all spatial relations.The models thus can be imagined as continuous sequences of 'spaces',i.e.models of M1.All the spaces of such a sequence have the same set of points but the

spatial relations can be different in each.The
'indices' by which the sequence is ordered are the
time-instances and it is required (via $\leqslant$) that
they in fact are ordered in a way isomorphic to
the order of real numbers.On the other hand space
and time are independent from each other to a
physically desirable extent.The relations concern=
ing time $\leqslant$, $\preccurlyeq$ and o are independent from space
by construction and the second component of d
-which describes time-distances-   is independent
from space by requirement (6) above.It should be
noted that models of M3 are logically much weaker
than all kinds of Riemannian space-times because
in M3 the spatial relations can undergo all kinds
of changes   -even non-continuous ones.

Now people in our fictitious culture of
course do not want to give up the simplifications
gained by coordinate systems.So they introduce
'space-time' coordinate systems.Such a CS for
$\langle P,T,\leqslant,\preccurlyeq,o,\underline{b},\equiv,d\rangle \in$ M3 is just a tuple
$\langle t^o,t^1,a_o,\ldots,a_3\rangle$ such that (1) $t^o,t^1 \in T, t^o \leqslant t^1$
and (2) for all $t,t' \in T$: $\langle a_o,\ldots,a_3\rangle$ is a CS for
$\langle P,\underline{b}_t,\equiv_t,d_{tt'},\rangle$. Such a CS yields a unique
'coordinatization' Υ for it can be proved that if
$x=\langle P,T,\leqslant,\preccurlyeq,o,\underline{b},\equiv,d\rangle \in$ M3 and $y=\langle t^o,t^1,a_o,\ldots,a_3\rangle$
is a CS for x then there exists a unique function
$\Upsilon:T \times P \to \mathbb{R}^4$ such that (1) $\Upsilon_{a_o}(t^o)=0$ and $\Upsilon_{a_o}(t^1)$
$=1$, (2) for all $t,t' \in T$ and all $a,b \in P$: $t \leqslant t' \leftrightarrow$
$\Upsilon_{a_o}(t) < \Upsilon_{a_o}(t')$ and $|\Upsilon_{a_o}(t)|=d_{ab}(t,t^o)$,
(3) for all $t \in T$ and $a,b \in P$: $\Upsilon_t(a_o)=0$ and $\Upsilon_t(a_i)=$
u_i (i=1,2,3) and $\|\Upsilon_t(a)-\Upsilon_t(b)\|=d_{tt}o(a,b)$,
(4) for all $t \in T$ and $a,b \in P$: $\Upsilon_a(t)=\Upsilon_b(t)$.Here,
$\Upsilon_a:T \to \mathbb{R}$ is defined by $\Upsilon_a(t)=\pi_1(\Upsilon(t,a))$,and
$\Upsilon_t:P \to \mathbb{R}^3$ is defined by $\Upsilon_t(a)=\langle\pi_2(\Upsilon(t,a)),$
$\pi_3(\Upsilon(t,a)),\pi_4(\Upsilon(t,a))\rangle$.Roughly, Υ consists of
just of the pair of the earlier Θ and s.Thus

with the help of a CS people are able to describe
motions by means of space-time coordinates.To
each relevant event of the motion they assign
four coordinates $\langle t, b_1, \ldots, b_3 \rangle$ relative to CS
$\langle t^0, t^1, a_0, \ldots, a_3 \rangle$, where t gives the 'time-distance'
to t^0 and $b_1, \ldots, b_3$ are the usual spatial coordi=
nates relative to the spatial coordinate system
$\langle a_0, \ldots, a_3 \rangle$.

Once again,independence from the choice of a
special CS is empirically found out as follows.
Whenever people investigate a concrete system
yielding a model of M3 and whenever y and y' are
two coordinate systems for x then the coordinati=
zations Υ and Υ' obtained by these coordinate
systems are connected by a space-time transfor=
mation,i.e.there are $b \in \mathbb{R}$, $\not{p} \in \mathbb{R}^3$ and a real,
orthogonal 3×3 matrix α such that $\Upsilon' =$
$\langle \pi_1(\Upsilon)+b, \alpha(\langle \pi_2(\Upsilon), \pi_3(\Upsilon), \pi_4(\Upsilon) \rangle) \rangle$. This
experience is made as long as both coordinate
systems y and y' in their spatial parts do not
move relative to each other.

Now my story ends with the following obser=
vation.If $x = \langle P, T, \lessdot, \lessdot, o, \underline{b}, \equiv, d \rangle \in$ M3,if y is a
CS for x and Υ is the unique coordinatization
$\Upsilon : T \times P \to \mathbb{R}^4$ given by y then s,defined by

$$s := \{ \langle p, \alpha, a, b, c \rangle / \exists t \in T(\langle t, p, \alpha, a, b, c \rangle \in \Upsilon) \}$$

is a function $s : P \times \mathbb{R} \to \mathbb{R}^3$.If s is smooth then
$\langle P, \mathbb{R}, s \rangle$ is a model of classical kinematics.And
the class of particle systems obtained from
$\langle P, \mathbb{R}, s \rangle$ by space-time transformations is identi=
cal with the class of all systems $\langle P, \mathbb{R}, s' \rangle$ ob=
tained from one and the same $x \in$ M3 by using
different coordinate systems y' yielding different
coordinatizations Υ',and consequently position
functions s'.If we write $s_{x,y}$ to indicate that
s comes from a model x of M3 via CS y in the way
described above then the last statement can be
formulated as follows.All and exactly all the
space-time transforms of system $\langle P, \mathbb{R}, s \rangle$ can be
obtained by fixing one space-time model $x \in$ M3 and

constructing all functions $s_{x,y}$ where y varies in
the class of all possible coordinate systems for
x.Expressed still differently we have: If x=
$\langle P,T,\leqslant,\measuredangle,o,\underline{b},\backsimeq,d\rangle \in M3$ and $M:=\{\langle P,\mathbb{R},s_{x,y}\rangle /$
y is a CS for x$\}$ then (1) any two members of M are
connected by a space-time transformation and (2)
if z $\in$ M and z' is obtained from z by a space-time
transformation then z' $\in$ M.

2. WHAT TO LEARN FROM THAT STORY?

1.) The story is not as fictituous as is suggested
by the headline of Sec.1.In fact,the historical
development on our planet led to Euclidean geo=
metry.Although its axiomatic form and complete=
ness of today was not available when CSs and CK
entered the scene there is no doubt that the
'invention' of coordinate systems at the times of
Descartes did presuppose the knowledge of Eucli=
dean geometry in its not yet perfect form.Also,
there is no historical doubt that clocks were
constructed before CK became important.So the
historical development from Euclidean geometry
and its CSs via clocks and its CSs to kinematics
roughly corresponds to that of our story.The only
point I would admit to be totally fictituous is my
treatment of time.I do not know of any historical
sources (i.e.say,from before 1850) in which a
foundational treatment of time is proposed which
tends into the direction of extensive systems.
(This is not so for space because if one examines
Euclid's axioms one can find a number of axioms
for extensive systems in his treatment of quanti=
ties,and the whole work is written in 'the spirit'
of extensive systems.) Concerning time,however,
foundational disputes have not ended up to now.
For different reasons there <u>is</u> no such commonly
accepted theory of time as is Euclidean geometry
for space.The development of CK as sketched above
may also be fictituous in one certain aspect to
which I will return in detail under 5.).But cer=
tainly it is correct in its rough logical struc=
ture which consists of bringing together inde=
pendent concepts of space and time.Remember that

I have in mind only <u>classical</u> kinematics.

2.) The story shows how the problem mentioned in the introduction can be solved.The problem was that,since the models of CK are invariant under a class of transformations much bigger than that of space-time transformations,there should be some explanation of why physicists are interested only in space-time transformations.In the story the solution is very simple.We just have to think of CK as being constructed on a space-time theory of the form M3 via coordinate systems.If this underlying theory is considered as a part of CK we obtain a theory in which space-time transformations of the position function in fact are the only ones admitted.More precisely,if we take models of CK to be entities of the form $x = \langle P,T, \lessdot , \prec , o , \underline{b}, \equiv ,d, s \rangle$ such that (1) $z := \langle P,T, \lessdot , \prec , o , \underline{b}, \equiv ,d \rangle \in M3$ and (2) there exists a CS y for x such that s is connected with x via y in the way described above (after the introduction of 'space-time' CSs) then (3) if $\langle z,s \rangle$ is such a model and s' is a space-time transform of s then $\langle z,s' \rangle$ is a model,too, and (4) structures obtained from $\langle z,s \rangle$ by differ=ent transformations are no models,i.e.if $\langle z,s \rangle$ and $\langle z,s' \rangle$ are models then s and s' are connected by a space-time transformation.[2] Intuitively,we can say that the class of space-time transforma=tions is exactly the invariance class of position functions relative to underlying space-time structures and their being connected with these by CSs.Still more roughly,invariance under space-time transformations is the correct invariance for position functions allowed by underlying space-time structures.It is only for practical reasons (reasons of simplicity) that this connec=tion to underlying theories is (systematically?) suppressed in physical treatments.

3.) The story served to underline my thesis formu=lated in the introduction.The origin of CK's in=variance under space-time transformations lies in observations   -as stressed in Sec.1-   of changes of CSs.Whenever systems are observed from differ=ent CSs their descriptions by position functions

are connected by space-time transformations.Thus
it is observed that the metrical relations   -ex=
pressed by the d of M3-   are invariant under
changing CSs which in turn yield space-time trans=
forms of the position functions.

    It might be objected that what I present here
as an empirical finding also might have been de=
duced from other assumptions.It would have been
possible as well,first,to observe that only cer=
tain changes of the CSs lead to identical results
with respect to metrical relations.In this way
people could have found a characterization of what
might be called admissible CSs.And from the concept
of admissible CSs space-time transformations might
be deduced.In principle I would be satisfied with
this story,too.For the only thing I want to stress
is that invariance has empirical roots.And the
alternative just mentioned certainly would have
its origins in experience,too.For experience
would be necessary to draw the distinction bet=
ween admissible and non-admissible CSs.But I
think the concept of an admissible CS -an ad=
missible CS is not the same as an inertial system-
did not play a central role in the actual develop=
ment,and,since this alternative story does not
yield more clarity,I stick to my story in Sec.1.

    Concerning the role of invariance the situ=
ation is a bit more difficult.The first aspect
mentioned in the introduction,namely that in=
variance shows a certain freedom of choice for CSs
is unproblematic.It even may seem trivial,al=
though it is not.It is certainly of practical im=
portance to know that some specified kinds of
transformations of the CS do not affect the system
under consideration.For instance,it is valuable to
know that we can approach our object until we see
it clear enough,or that we can go around,say,a
picture in order to observe it from the front side
and not from a very uninformative point of view
situated on the wall were the picture is hanging.

    But there is a second component in the role
of invariance which,I must confess,has not been
illuminated by my story.This role can be seen by
passing from passive transformations to active
transformations.Up to now we have always been

thinking of passive transformations in the fo=
llowing sense.It was assumed that the actual
physical system under consideration remained un=
affected while the CS   -in its real representa=
tion given by $\langle t^{o},t^{1},a_{o},...,a_{3}\rangle$-   was thought to
be changed.We obtain active transformations if the
real CS $\langle t^{o},t^{1},a_{o},...,a_{3}\rangle$ is left unchanged and the
physical system under consideration from the point
of view of this CS is changed.Such a change can
consist in bringing the physical system to another
place or to 'start it' at a later time.As an ex=
ample consider a pendulum swinging in a laboratory.
If we pass from the CS given by the laboratory's
walls to a CS installed immediately in front of
the pendulum in form of some iron frame we have a
passive transformation.If we take the pendulum,
bring it to another corner of the laboratory,and
there start it swinging again we have an active
transformation.

Now although invariance primarily has come up
with passive transformations its 'discovery' cer=
tainly tempts to investigate the reverse situation
of active transformations.And in fact there are
quite a number of concrete situations in which
space-time transformations can be executed acti=
vely without much change of the system.This is
true,for instance,for systems of solid bodies.It
is also true for motions which in a certain sense
can be controlled by humans,e.g.pendulae,cars and
a wide class of technical applications.For this
big class of phenomena with its overwhelming
practical importance there is an empirical in=
variance under active (not too extreme) space-
time transformations.Since this kind of invariance
in its theoretical description does not differ
from the passive case it seems not unfair to take
it as the second component of the role of invari=
ance in CK.

4.) The story should please philosophers of
science for two reasons.First,it contains a lo=
gical reconstruction of the underlying theories of
kinematics and therefore it makes explicit how
classical mechanics is based or can be based on

theories of space and time.By introducing suitable
intertheoretic relations   -as for instance theo=
retization (compare Balzer (1978))-   we can build
a small hierarchy of theories such that classical
mechanics is the top element of this hierarchy
and its basic elements are theories containing
only qualitative basic terms.So we have one
possible way of depicting how quantitative theo=
ries rest on qualitative,'proto'-physical theories.
      Second,the story throws some light on the
connection between mechanics and measurement of
space and time.For it contains conditions under
which unique distance-values for space and time-
distances can be obtained from qualitative oper=
ations   -as expressed by the qualitative terms
of M1 and M2.And it shows how these distance
values   -without essential additional assumptions-
go into kinematical descriptions.I do not want to
say,of course,that Euclidean geometry or extensive
systems describe real measurements.But it seems
plausible to expect descriptions of actual methods
of measurement for space and time to yield sub-
structures of models of geometry or extensive
systems.Thus,the connection between actual measure=
ment and mechanics can be worked out by studying
sub-structures and their intertheoretic relations
'inside' the hierarchy mentioned above.

5.) A last point we can extract from the story is
this.Space-time can be described in a way that
makes no difference between points of space and
particles.In order to understand precisely what I
mean by this it is necessary to describe the
alternative in which points and particles are
treated differently.In this alternative approach
spatial distances among points   -in contrast to
particles-   are required to remain fixed in time.
That is,the distances between any two points a,b
are the same at all times t.In the language of M3
this amounts to requiring that for all $a,b \in P$ and
all $t,t',t_1,t_2 \in T: d_{tt'}(a,b) = d_{t_1 t_2}(a,b)$.Intuiti=

vely,this means that the spatial distances are
independent of time.If the models in M3 are re=
quired to satisfy this additional condition then

space-time becomes 'rigid'.By this I mean that the
models of such a theory do not allow for motions.
Remember that in M3 motion is described by change
of spatial distances relative to time-distances.
If the spatial distances are not allowed to change
no motion is possible.These rigid models can be
imagined as a sequence of models of geometry such
that in any two spaces of the sequence all points
are 'at the same place'.The visual picture of this
situation is given by $\mathbb{R}^4$ (or rather by $\mathbb{R}^3$ with
a two-dimensional space).For each $\alpha \in \mathbb{R}$ the cuts
$\{\alpha\} \times \mathbb{R}^3$ are models of geometry,and all these
cuts are neatly built one upon another such that
'vertically' there are no differences if we go
through the cuts.
        Kinematics in this approach is treated by
introducing into such a rigid space-time finitely
many particles moving around.This can be done,e.g.
by introducing another set Q of objects -namely
particles- and a function $i:Q \times T \rightarrow P$ assigning to
each particle and each point of time a 'point of
space',namely the point of space where the parti=
cle is at that time.
        This way of reconstructing kinematics seems
to be closer to physicists' present day thinking.
It is just nice to have a neat,rigid space-time
sharply to be distinguished from things moving
around.Our considerations about invariance would
remain essentially the same -perhaps with some
slight complications- when we use this kind of
reconstruction for CK.What we can learn from the
story then is that this alternative way of re=
constructing CK is not the only one.There are
other possibilities,one of which consists in
dropping the distinction between points of space
and particles.[5] In models of M3 points are allowed
to change their distances during the flow of time.
That is,there can be $t,t',t_1,t_2 \in T$ and $a,b \in P$ such
that $d_{tt'}(a,b) \neq d_{t_1 t_2}(a,b)$.The disadvantage of this
approach,in the physicist's view,consists of
giving up a distinction,namely that between points
of space and particles which seems intuitively
clear.The question whether this intuition is

sound I leave for Sec.3.

3. SOME POLEMIC REMARKS

I want to conclude with some more general remarks
on classical space-time and present day physics
which are likely to be felt as polemics because
their origin is far away from present day para=
digms -in Kuhn's sense- of physics.
   Let me start by pointing out that today space-
time physicists throughout use the language of
manifolds and of representations by transforma=
tion groups.My treatment in terms of 'old fashi=
oned' logics comes really from an other world.I
agree that modern representations of space-time
are framed in a way which allows to depict vari=
ous kinds of older and weaker theories in a way
that is felt to be very elegant.As far as physics
is concerned such a treatment is all right be=
cause it shows how older theories are related to
and how they can contribute to a better under=
standing of the most recent theories.From the
point of view of philosophy of science or history
of science,however,this advantage may turn into a
disadvantage.For what is gained by modernizing old
theories -i.e.reformulate them in a modern frame-
may be lost in understanding these older theories
in an historically adequate way.It is of course
difficult to say precisely what is an historically
adequate representation.But twenty years of dis=
cussion in the philosophy of science should have
taught us that there is not only a possibility
but even a certain probability of incommensurable
world views.To stick to most modern reformulations
of old theories as historically adequate therefore
bears the risk to become rather one-sided -some
even call it blind- in certain respects.
   I do not want seriously to entertain the idea
that a logical frame as used in Sec.1 could be
used as an alternative to prevailing modes of
description.But I do so only because obviously no=
body has really thought about this possibility
and because I have no precise elaboration.I do
want,however,to say one more word about histori=
cal adequacy.In the present example of space-

time we can find a nice example of meaning-vari=
ance   -if not incommensurability-   ,namely the
term 'event'.Certainly this term has existed in
Newton's times.But it was no technical term of
mechanics.People were talking about events only in
ordinary language as they still do today.Only after
Minkowski's treatment of special relativity events
became a technical term,in fact a <u>central</u> object-
term of space-time theories.For now events are the
basic objects out of which models consist.This
role in classical theories was played by points
(or particles) and points of time.It was essential
to the notions of point and instant that both were
independent from each other.Although the notion
of a classical point of time is somehow unclear
it can be hold for sure that it is different from
our new concept of event.So the shift in the
meaning of 'event' is that from an ordinary lang=
uage term in which reference to space and time
was not necessarily included to a technical term
not only containing the concepts of space and
time but essentially containing their depending
on each other.[6] In the logical reformulation of
Sec.1 events are not necessary as basic objects.
So it has more chance to be historically adequate
than modern physical formulations.This point be=
comes even more drastic if we consider,say,Rie=
mannian connections instead of events.Such things
did not exist at the old times.Similarly,I would
claim that to apply a four-dimensional frame is
adequate only after Minkowski.

But in addition to this difference in lang=
uage my treatment differs in content,as already
indicated.For I have treated space-time not as
rigid   -allowing for motions 'in' space-time-
whereas the tendency seems to be to distinguish
space-time and kinematics and to introduce moving
particles only after a rigid or at least in some
sense complete space-time is at hand.[7]

Now besides having demonstrated that this
must not necessarily be so I have three arguments
in favour of non-rigid space-times.First,I have
Ockham's razor.With this I cut off all terms
which are not really necessary.And if the dis=
tinction between points and particles is not

really necessary   -which is <u>proved</u> by showing
how to do without it in Sec.1-   one of these two
is superfluous.Second,the 'rigid space-time view'
has the problem of giving a physical meaning to
the function i mentioned in 5.) of Sec.2.What does
'i(q,t)=p' mean? I do not want to say that it is
impossible to give reasonable meaning to this
statement.But it seems rather difficult to do so.
The problem is to give meaning to the 'at point
p' part independently from the fact that q at t is
there.Third,and connected with the last remark,I
can ask: What <u>are</u> points of space if they are not
materialized or not materializable by particles?
Do they exist? And how? I think,on a naive
approach to the alternative "There is a clear
distinction between points and particles" and
"There is no sense to talk about points not
materializable by particles" we need not nece=
ssarily prefer the first one.But even if we pre=
fer the first alternative the second one still
provides a problem.This problem is not solved up
to now.
      Finally,in today's space-time theories as
well as in other physical theories there is a
tendency to seperate mathematics from physics
which brings me to a point of theory-formation in
general.What is the role of mathematics in empi=
rical theories? A first move to answer this quest=
ion is to look on the historical development and
to say that mathematics   -at least $\mathbb{R}^4$ and re=
lated stuff- is the outcome of physical theories
and therefore a kind of 'physical image' of some
aspects of the world.But today those mathematical
structures are used in quite different areas as
well and the development of non-Euclidean geome=
tries has shown that there is not one single true
space-time structure.So one is led to admit a
variety of mathematical 'images'.The result has
been a total separation of mathematical formalism
and 'physical reality',as expressed e.g.in Ludwig
(1978).Such a separation,however,necessitates the
introduction of additional correspondence prin=
ciples relating the real and the mathematical
parts of a theory to each other.Now what is the
status of such correspondence principles? No clear

examples can be found in the literature.No clear
interpretation of such principles is available
yet.In view of the fact that these difficulties
come from the want for separation of mathematics
from physics   -and only from there-   the alter=
native of trying to diminish the need for such
rules by diminishing the separation of mathematics
from physics seems to become more and more
attractive.

Seminar für Philosophie,Logik und Wissenschafts=
theorie,Universität München

NOTES

0) I am indebted to A.Kamlah for helpful discus=
   sions.
1) This notion of kinematics differs from that of
   McKinsey et al.(1953) in two respects.First,
   time is represented by the whole set of real
   numbers and not by an open interval.This is no
   real difference because both these things are
   isomorphic   -at least topologically.Second,
   the set P of points or particles is not requir=
   ed to be finite.This difference is essential
   and the reason for it is discussed in 5.) of
   Sec.2.
2) This is not quite true because dilatations
   cannot be ruled out in this way.However,I will
   neglect this point since dilatations have a
   clear physical meaning   -freedom of choice of
   the unit of measurement-   which can be sharply
   distinguished from that of space-time trans=
   formations.
3) Of course 'true' here does not mean 'true in
   the sense of the usual inductive definition in
   all of the domain'.It just means that 'suffi=
   ciently many' instances of sufficiently 'de=
   quantified' forms of the axioms are true.In
   order to avoid complications axiom (13) of
   Tarski (1959) has to be replaced by the corres=
   ponding second order formula on page 18 l.c.
4) Since the betweenness relation is not a rela=
   tion among rods some interpretation is nece=

ssary here.babc can be interpreted as saying that the rod with end points a,c is just the concatenation of the rods with end points a,b and b,c.

5) It is our neglect of this distinction that forces us to allow P to be infinite (compare footnote 1)).

6) This example is discussed in some more detail in Balzer (1978b).

7) In the case of general relativity this leads to a subtle situation.For there,space-time is not rigid at all.Rather it is 'formed' by the material particles.But it must be stressed that this is only a kind of 'surface' space-time because general relativity logically contains what I call a rigid and complete space-time.Such a space-time is implicit in the very mathematical formulation in the form of e.g. $\mathbb{R}^4$ which is needed to define 4-dimensional differantiable manifolds.

REFERENCES

Balzer,W.: 1978, Empirische Geometrie und Raum-Zeit-Theorie in mengentheoretischer Darstellung, Kronberg i.Ts.

Balzer,W.: 1978b, 'Incommensurability and Reduction',Acta Philosophica Fennica,Vol.37,Nos.2-4

Krantz,D.H.,Luce,R.D.,Suppes,P.and Tversky,A.: 1971, Foundations of Measurement,New York-London

Ludwig,G.: 1978, Die Grundstrukturen einer physikalischen Theorie,Berlin-Heidelberg-New York

McKinsey,J.C.C.,Sugar,A.C.and Suppes,P.: 1953, 'Axiomatic Foundations of Classical Particle Mechanics',Journal of Rational Mechanics and Analysis II

Tarski,A.: 1959, 'What is Elementary Geometry?', in: Henkin,Suppes,Tarski (eds.): The Axiomatic Method,Amsterdam

Andreas Kamlah

THE SIGNIFICANCE OF PHYSICAL INVARIANCE
PRINCIPLES FOR THE MEASUREMENT OF SPACE - TIME
QUANTITIES

1. INTRODUCTION

In the second half of the last and in the first decade of
our century the group theoretical approach to geometry has
been very popular. It was H.v. Helmholtz who first attempted
to derive the geometrical axioms from postulates describing
the possible motions of rigid bodies. Helmholtz' ideas have
been developed further by S. Lie who showed how Euclidean and
some other geometries may be characterized by certain groups of
differentiable transformations in space. Groups however have
not only be considered as mathematical tools useful for geometry, they seemed also to be a link between geometry and
physics, between the mathematical theory of space and the
real world. The significance of groups for spatial measurements is already inherent in v. Helmholtz' idea to characterize geometry by the motions of rigid bodies, which are
usually applied for measurements of length, measuring rods
being just special cases of them. Thus Poincaré writes:
"The object of geometry is the study of a particular 'group'
...... from among all possible groups we must choose one
that will be standard, so to speak, to which we shall refer
natural phenomena. - Experiment guides us in this choice,
which it does not impose to us." (1952, S. 70) We shall not
deal here with Poincaré's own standpoint, what is important
for us is only that natural phenomena are put into a relation with the group, which is characteristic for the
accepted geometry. Also in B. Russels dissertation and for
many other natural philosophers of that time the group theoretical approach is used for studying physical geometry.
The advent of general theory of relativity however stopped
the whole discussion on the significance of groups for physical geometry since groups play no important role in this
new theory of space-time.
    This loss of interest for geometrical groups was at
least premature. Though it has to be conceded that groups
can never be the only link between geometry and the real

D. Mayr and G. Süssmann (eds.), Space, Time, and Mechanics, 171–194.

world, they play an important role for semantics of geometrical concepts after all.

It is the role which we want to show up in our paper
together with some of its implications. At first we want to
show that the symmetries of nature, the invariance of all
physical laws are independent of the language of physics.
Physics may be reformulated in any different language without affecting the invariances. Later on we shall investigate
the role played by these invariances for measuring procedures
of length, duration of time, or space-time distance.

We shall not be concerned here with a group theoretical
axiomatization of spatial or space-time geometry as are for
instance H.J. Schmidt or D. Mayr in this volume. Our goal is
different. While H.J. Schmidt presupposes the existence of
rigid bodies, we start by questioning it and by asking how
rigid bodies might be defined. Thus groups in space, time
and space-time are not used by us to characterize geometry,
the rigid bodies being given, but rather to characterize the
rigid bodies and ideal clocks themselves. Therefore we start
our discussion in the next section by studying proposed definitions of the ideal rigid body.

2. THE PROBLEM OF RIGID BODIES

According to a view on physical geometry commonly accepted
by operationalists geometric properties of space are determined by measuring rods which do not change their form by
definition, when they are transported. These rods define the
congruence relation, in terms of which the second fundamental relation of geometry, the betweennes relation may be defined (see W.Balzer, A.Kamlah 1980). Thus the congruence relation may be considered as the "structure term" of geometry
and is the only geometrical relation, which needs a physical
interpretation. Such an interpretation could easily be
established if there were a class of ideal rigid rods, which
behave consistently, for which pairs of marks on two rods
which coincide at one place in space do this also at any
other place, where they are brought together. Unfortunately
in our real world there is no such class, all solid bodies
may be deformed by some influences such as elastic forces.
But the rigid body may be reconstructed from solid bodies if
these influences are taken into account and corrected for.
Now the question arises: Is it possible - as H. Reichenbach believed - to give a prescription for these corrections
without referring to any special physical theory such as

electrodynamics? Reichenbach tried to divide the deforming forces into differential forces which are detectable since they work differently on different solid bodies, and into universal forces which influence all solid bodies in the same way and are therefore undetectable in regions of space much smaller than the radii of inner curvature. Reichenbachs prescription is now: Correct for differential forces and consider universal forces as nonexistent.

As Hilary Putnam has shown (H. Putnam 1963) Reichenbachs distinction is not without ambiguities but depends on the language used for the formulation of physical theories. Putnam introduced nonstandard physical fields which might be used for the description of our physical world instead of our usual ones and showed that certain universal forces become differential in this new description (1963, p.247-255; 1979, p.124-129).

Putnams criticism is much in favour of P. Duhems conventionalism or holism, according to which geometry has a meaning only as an element in a set of consistent physical theories, in terms of which a rigid body may be defined. The most famous holist has been A. Einstein in spite of the admiration given to him by the operationalists. He puts his point already very clearly in his essay "Geometry and Physics" and later in an invented dialogue between Reichenbach and Poincaré. There he lets Poincaré say

"In gaining the ... definition [of the rigid body] improved by yourself you have made use of physical laws, the formulation of which presupposes (in this case) Euclidean geometry. The verification [of Geometry] , of which you have spoken, refers, therefore, not merely to geometry but to the entire system of physical laws which constitute its foundation. An examination of geometry by itself is consequently not thinkable" (A. Einstein 1949, p.677).

In the last twenty years Einsteins holism, the position that only the whole of physics and geometry is empirically testable, has been much more fashionable than operationalism. There are indead strong arguments in favour of Holism or the Duhem-Quine thesis as A. Grünbaum calls this standpoint, and this view seems indeed to be very suggestive. But nobody has really proved until today, that there is indeed no way to give an operational definition of geometrical terms which is immune against the holistic objections. Therefore there seems to be still a chance for a compromise between holism and operationalism. In this paper we shall defend the claim that the physical invariance principles make possible

a partial empirical test of geometry and a partial definition of space-time concepts, which are both independent of
the remaining physical theories and of the language used for
their formulation. If we want to understand how invariance
principles might become independently testable we have to
remind ourselves of the way how Duhem has put forward his
holistic argument. For Duhem geometry G is a subtheory of
physics, a part of a logical conjunction of all physical
theories $P_1 \ldots P_n$. If we want to derive an empirical consequence E from geometry G we have to use some physical hypotheses $P_1 \ldots P_n$ as additional premises, since no proposition
about measuring instruments follows from G alone. We therefore have
(1) $G \wedge P_1 \wedge \ldots \wedge P_k \longrightarrow E$
If our experiments tell us now, that E is the case, we are
led to the conclusion, that at least one of the theories G,
$P_1, \ldots, P_k$ is false. But unfortunately we do not know which
one. Therefore geometry G is not testable independently of
the assumed theories $P_1, \ldots, P_k$. In order to undermine Duhems
argument we shall point out, that not all physical propositions
are statements in the object language of physics. There are
also statements in physics which have the status of meta-pro-
positions, of propositions about physics, and the most important principles of physics belong to this class of propositions. Take for example the principle of indeterminism, in
the formulation: There are some irreducibly indeterminate
physical laws. This principle is clearly no sentence of the
object language; it talks about physical laws. It is corrobor-
rated if at least one irreducibly indeterministic physical
law is sufficiently well confirmed; (whatever "irreducible"
means here). Duhems deductivist scheme of falsification however is not applicable to a principle of this type.

Invariance principles are also metapropositions. They
hold all physical laws to be invariant under a certain group
of transformations. Thus a certain invariance or else a certain geometric structure is a property of a complete set of
physical theories and not one physical theory amongst others.
If we can show that this property is independent of the
linguistic formulation of these theories, invariances of geometrical symmetries of space or space-time are immune against
conventionalistic reformulations of physics. Therefore if
an evidence E confirmes a conjunction of theories $P_1 \wedge \ldots \wedge P_n$
or an empirically equivalent conjunction $P'_1 \wedge \ldots \wedge P'_m$ it
will also confirm their geometric symmetry which is common to
$P_1 \wedge \ldots \wedge P_n$ and $P'_1 \wedge \ldots \wedge P'_n$. In the next section we show,

that a language independent characterization of physical invariance is possible.

If we have shown that important axioms of geometry are
independently testable, we still have to discuss the question
how far the spatial distance function or congruence relation,
duration of time, and space-time distance may be determined
by the invariance principles alone. For such determination
we need of course an additional requirement. By demanding
that the spatial, temporal, or space-time distance function
is itself invariant under the physical invariance transformations, I am led in another paper (A. Kamlah 1979) to the
result that spatial distances are not uniquely defined by
the mentioned requirement, temporal distances are, if a very
general principle is added and the same in true for space-
time distances. In any case however the important role of
the invariance principles for the definition of space-time
concepts becomes evident.

3. INVARIANCE

We now have to make clear in which respect geometry is a
statement about physics. More exactly: The statement that
the geometry of a possible world belongs to a special symmetry
class is a statement about the natural laws in this world.
Geometries may be classified according to Lie-groups of
transformations which leave invariant all laws of physics.
The invariance of a theory in respect to a certain mapping
of all its state descriptions into others is a property
which cannot be changed by translation into any other theoretical language. So if it can be brought into correspondence
with a symmetry class, to which the geometry belongs, we have
found a partial characterization of the geometry of a world
which is immune against conventionalistic ("Duhemistic" in
Grünbaums sense) critism.

We have now to specify our claim that physical invariances are language independent. At first sight this seems
to be quite a trivial claim. For the discussion of this thesis
we want to use J.D. Sneeds model theoretic formalism.

According to Sneed a physical description may always be
considered as involving some sets of individuals $M_1,\ldots,M_k$
and some predicates (relations or functions) $P_1,\ldots,P_m$ which
are defined on these sets. The predicates $P_1,\ldots,P_m$ "talk
about" the objects, which are the elements of $M_1,\ldots,M_k$. A
series of possible states in time of a physical system or
in other terms a physical process is then given as an ordered

set $<M_1,...,M_k, P_1,...,P_m>$, which designates a "possible model" in Sneeds terminology for the physical theory. Let us call the set of these possible models M_{pot}. An alternative description will then lead to a second set M'_{pot} of possible models $< M'_1,...,M'_{k'}, P'_1,...,P'_{m'} >$ and a translation of description into another may thus be represented by a map f from M_{pot} onto M'_{pot}

(2) $f : M_{pot} \longrightarrow M'_{pot}$.

Next we have to characterize physical invariance in the model theoretic scheme. According to Sneed a physical theory is represented by the set of possible models, which are obeying the laws of this theory or more briefly by the set of models M of this theory.

(3) $M \subseteq M_{pot}$.

The invariance transformations $t \in G_T$ are maps from M_{pot} on itself, homomorphisms of M.

(4)   For all $t \in G_T$ $(t : M_{pot} \longrightarrow M_{pot})$.

The Theory T is then said to be invariant under a group of transformations G if

(5)   For all $t \in G$ $(x \in M \longrightarrow t\,x \in M)$.

This is still not the whole truth about invariance of physical laws in Sneeds frame of theory construction. Sneed's concept of a physical theory is indeed much more complicated. We want here to use a simplified scheme which was already applied by Adams. For Adams a theory consists of a set of models M an additionally of a set I of intended applications of the theory. We may take as a simple example the theory of ideal gases. Here M_{pot} may denote a set of descriptions of portions of matter. $I \subseteq M_{pot}$ denotes the set ot those descriptions M_{pot} which are considered as "ideal gases" if the relevant physical quantities are measured in the usual way. Finally the elements of M are those descriptions which satisfy the equation of Avogadro $p \cdot V = R \cdot T$.

Thus the theory of ideal gases does not claim that all descriptions of probes of matter obey Avogadros law; this is only claimed for the intended applications, for those substances which are held to be approximately ideal gases. The set I therefore seems to be a very important feature of an adequate concept of physical theories. It accounts for the fact, that physicists seldomly claim a theory to be universally valid. In most cases the validity of the theory is only claimed for a limited area of intended applications.

The author has now proposed to split up the set I into two sets I_R and I_s, such that $I_R \cap I_s = I$. (A. Kamlah 1976, p. 355). Here I_s denotes the set of those $x \in M_{pot}$ which are the descriptions of real physical processes obtained by the measurement procedures for the quantities and concepts of the theory. I_R is the set of those descriptions of physical processes which satisfy certain conditions under which the theory will be a good theory . So for example the condition $v \leq c$ (c = velocity of light) has to be presupposed for any application of classical nonrelativistic mechanics.

The decomposition of I into I_R and I_s becomes important, if one wants to express the invariance of a theory under a group of transformations G. For not only the set M has to be invariant, it seems reasonable to demand that also the condition of applicability i.e. the set I_R has to be invariant under all $t \in G$. For I_s however such a requirement would be completely inadequate. I_s is the set of those possible models $x \in M_{pot}$ which are descriptions of physical processes in the real world, if the physical quantities in question are measured in the prescribed way. If I_s in celestial mechanics would be taken to be translational invariant, it would demand, that at any place in the universe there is a solar planetary system, provided there is one such system at one place. In other terms, if I_s would be translational invariant, matter had to be distributed homogeneously in the universe. We cannot accept such an implication. We want to express the invariance of physical laws without requiring that the world as such is invariant. Therefore we may conclude that a physical theory $< M_{pot}, M, I_R, I_s >$ is invariant under a group of transformations G if

for all $x \in M_{pot}$ and for all $t \in G$:

(6) $x \in I_R \longrightarrow tx \in I_R$ and $x \in M \longrightarrow tx \in M$.

So far we have given a rational reconstruction of the invariance of physical theory under a group of transformations. We have now to study what happens to a physical invariance principle if the theory is translated from one description into another description. Let us assume that we have two description of the same theory. In the first the theory

"talks about" the basic sets $M_1,\ldots,M_k$, in the second about
the sets of individuals $M'_1,\ldots,M'_{k'}$. The first description
applies to the sets $M_1,\ldots,M_k$ the predicates $P_1,\ldots,P_{m'}$ the
second to $M'_1,\ldots,M'_{k'}$ the predicates $P'_1,\ldots,P'_{m'}$.
So the sets of possible models are

$$M_{pot} = <M_1,\ldots,M_k, P_1,\ldots,P_m >$$

and $$M'_{pot} = <M'_1,\ldots,M'_k, P'_1,\ldots,P'_{m'} >$$

We may call the sets of models of the theory M and M'

(7) $$M \subseteq M_{pot} \quad \text{and} \quad M' \subseteq M'_{pot}$$

and the sets of intended applications $I = I_R \cap I_s$ and
$I' = I'_R \cap I'_s$ respectively, which are typified analogously.
    A translation from one description into the other may
be characterized as a bijection f from M_{pot} to M'_{pot}

(8) $$f : M_{pot} \longrightarrow M'_{pot}$$

        $$f^{-1}: M'_{pot} \longrightarrow M_{pot}$$

which satisfies the following conditions:

(9)   I $x \in M \longrightarrow f\ x \in M'$
        II $x \in I_R \longrightarrow f\ x \in I'_R$
        III $x \in I_s \longrightarrow f\ x \in I'_s$

In order to show that the theory must have the same invariance
properties in both representations, we first introduce the
invariance transformations $t \in G$ and $t' \in G'$. G and G' are
the groups of invariance transformations in both cases. We
may now prove the following theorem.
T 1: Is G a group of invariance transformations of a theory
$< M_{pot}, M, I_R, I_s >$ the translation of the first theory obtained by the bijection
$f: M_{pot} \longrightarrow M'_{pot}$, there will be a Group G' of transformations isomorphic to G, which is an invariance group of
$< M'_{pot}, M', I'_R, I'_s >$ such that

        $$t \in G \longrightarrow f\ t\ f^{-1} \in G' .$$

The proof is simple. G' may be constructed as follows
$$G' = \{x\,|\,\bigvee t\ (t \in G \wedge f\ t\ f^{-1} = x)\}$$

G' is a set of bijections form M'_{pot} to M'_{pot} since f^{-1} operates from M'_{pot} into M_{pot}, t from M_{pot} into M_{pot} and f from M_{pot} into M'_{pot}. All $t' \in G'$ map M' and I_R on themselves. From

$$\text{for all} \quad t \in G \; (x \in M \longleftrightarrow tx \in M)$$
$$x \in M \longleftrightarrow fx \in M'$$
$$x \in M' \longleftrightarrow f^{-1} x \in M$$

and from the construction of G' follows that for all $t' \in G'$ holds: $y \in M' \longleftrightarrow t'y \in M'$. For I'_R the proof runs in the same way as for M'. Now we have to show that G and G' are isomorphic. Let t_1, t_2 and t_3 be elements of G and $t_1' = f \, t_1 \, f^{-1}$, $t_2' = f \, t_2 \, f^{-1}$ and $t_3' = f \, t_3 f^{-1}$ the corresponding elements of G'. Then

$$t_1 t_2 = t_3 \longleftrightarrow t_1' t_2' = t_3'$$

since

$$t_1' \, t_2' = f \, t_1 \, f^{-1} \, f \, t_2 f^{-1} = f \, t_1 \, t_2 \, f^{-1} = f \, t_3 f^{-1} = t_3'$$

This means that f maps the group theoretical multiplication operation of G into the corresponding multiplication operation of G'. Since this operation is the only nomological operation in group theory, both groups are indeed isomorphic. Thus the proof of T1 has been established.

A corrollary of T1 is :

T2   The group G_T of all invariance transformations of $\langle M_{pot}, M, I_R, I_s \rangle$ is isomorphic to the group G_T' of all invariance transformations of $\langle M'_{pot}, M', I'_R, I'_s \rangle$.
The proof is a trivial application of T1.

From T1 one might conclude, that we have now attained our goal. We have the invariance principle of a physical theory shown to be a metastatement independent of the object language. For T1 says that the group of invariance transformations has the same structure in both representations of the theory. This result however is only valid under the given premises. We presupposed that the reformulation of a theory in a different language is in any case given by a bijection f from M_{pot} to M'_{pot}. This needs not to be the case. If one form of the theory is Maxwells electrodynamics formulated in terms of H and E, the magnetic and electric field, and the other form of the theory is electrodynamics formulated in terms of the vector potential A and the scalar potential ϕ, everybody knows that in the latter representation the theory has additional invariance properties, it is "gauge" invariant, what is not the case for the former. But it is also known that this gauge invariance is of no empirical significance. In the

second representation we have simply introduced additional
degrees of freedom which do not correspond to any measureable
quantity. Therefore in the second representation there are
sets of possible models which may be transformed into one-
another by empirically undetectable transformations. These
maps are empirically undetectable, since all possible models
of those sets, the elements of which differ only in the additional or spurious degrees of freedoms, are empirically equivalent.

In the next section we shall separate the physically
significant transformations from the gauge transformations,
and shall show that the physically or empirically significant
invariance group is independent from the linguistic formulation of the theory.

4. ELIMINATION OF THE GAUGE TRANSFORMATIONS

This section will be of a more formal character than the
preceeding and the following ones. The reader therefore may
leave it out, if he does not like to follow the sequence of
its axioms, definitions and theorems and if he prefers simply
to accept the results formulated in theorems T10-12. Nevertheless these theorems are of central importance for the
whole paper, and therefore it may be justified to prove them
in a more mathematical fashion instead of making them merely
plausible in an intuitive way. At the end of the last section
we have seen that a language independent rational reconstruction of physical invariance is not yet obtained if the group
G_T in the sense of theorem T2 is simply considered as the
symmetry group in question. The group may still contain gauge
transformations, which are simply mapping different formulations of the theory on each other. What we need, is something as the group of empirically significant transformations.
For this purpose we have to introduce an additional concept,
a relation of empirical equivalence between two different
possible models of the theory. We have assumed, that the two
descriptions of physical processes by the elements of M_{pot}
and M'_{pot} refer to the same states of affair. Therefore there
must be a relation
$$Ao \qquad \underset{t}{\sim} \subset M_{pot} \times M'_{pot}$$
such that $x \underset{t}{\sim} x'$ says that $x \in M_{pot}$ and $x' \in M'_{pot}$

refer to the same physical process, and can be considered
as translations of each other. The relation $\underset{t}{\sim}$ replaces the
bijection f which we have used initially in our discussion.

The relation $\underset{t}{\sim}$ is a two-place relation like f but has to be no function; no argument of $\underset{t}{\sim}$ has to be uniquely determined by the other. If there would be no such relation $\underset{t}{\sim}$ it would not be possible at all to say, that the first description (by the elements of M_{pot}) and the second description (by the elements of M'_{pot}) talk about the same states of affair in different languages. The relation $\underset{t}{\sim}$ has to satisfy the following axioms:

A1a $\qquad x \in M_{pot} \quad \longrightarrow \bigvee x' \ (x' \in M'_{pot} \wedge x \underset{t}{\sim} x')$

A1b $\qquad x' \in M'_{pot} \quad \longrightarrow \bigvee x' \ (x \in M_{pot} \wedge x \underset{t}{\sim} x')$

For any $x \in M_{pot}$ or $x' \in M'_{pot}$ there has to be a translation into the other description.

A2 $\qquad x \underset{t}{\sim} x' \wedge y \underset{t}{\sim} x' \wedge y \underset{t}{\sim} y' \longrightarrow x' \underset{t}{\sim} y'$

If x and y are both translations of x' and y is also a translation of y', then y' must be a translation of x too.

A3a $\qquad x \underset{t}{\sim} x' \longrightarrow x \in M \longleftrightarrow x' \in M'$

A3b $\qquad x \underset{t}{\sim} x' \longrightarrow x \in I_R \longleftrightarrow x' \in I'_R$

A3c $\qquad x \underset{t}{\sim} x' \longrightarrow x \in I_s \longleftrightarrow x' \in I'_s$

The sets M, I_R, I_s and M', I'_R, I'_s have to be mapped onto each other by the translation. (This is a rather strong condition of equivalence of two theories, which could be replaced by a weaker one. Though this more adequate account would lead to the same results, it would make our discussion more complicated, and is therefore left out of consideration here).

We may now easily define two relations of empirical equivalence for possible models of both representation of the theory

D1a $\qquad x \underset{e}{\sim} y \leftrightarrow \bigvee z' \ (z' \in M'_{pot} \wedge x \underset{t}{\sim} z' \wedge y \underset{t}{\sim} z')$

D1b $\qquad x' \underset{e}{\sim} y' \leftrightarrow \bigvee z \ (z \in M_{pot} \wedge z \underset{t}{\sim} x' \wedge z \underset{t}{\sim} y')$

From the definition it is clear that $\underset{e}{\sim}$ and $\underset{e'}{\sim}$ are of the following type:

$$\underset{e}{\sim} \subset M_{pot} \times M_{pot}; \ \underset{e'}{\sim} \subset M'_{pot} \times M'_{pot}$$

The intuitive idea underlying these two definitions is that two possible models of a theory must be empirically equivalent if they can be translated into the same model of a different description and vice versa.

We may easily show that $\underset{e}{\sim}$ and $\underset{e'}{\sim}$ are equivalence relations.

T3 $\underset{e}{\sim}$ and $\underset{e'}{\sim}$ are transitive, symmetric and reflexive.

Proof: The proof of symmetry and reflexivity is trivial.

The transitivity of $\underset{e}{\sim}$ we shall prove indirectly. For $\underset{e'}{\sim}$ the proof will be analogous. Let us assume that $\underset{e}{\sim}$ is not transitive. Then on behalf of D1 we have $x_{1t}\underset{}{\sim}x_1'$, $x_{2t}\underset{}{\sim}x_1'$

$x_{2t}\underset{}{\sim}x_1'$, $x_{2t}\underset{}{\sim}x_2'$, $x_{3t}\underset{}{\sim}x_2'$, but for any $x_3' \in M_{pot}$ either $x_{1t}\underset{}{\sim}x_3'$

or $x_{3t}\underset{}{\sim}x_3'$ is wrong. Let us assume that the first is the case, that $x_{1t}\underset{}{\sim}x_3'$ holds, but $x_{3t}\underset{}{\sim}x_3'$ is wrong. We may now conclude from $x_{1t}\underset{}{\sim}x_1'$, $x_{2t}\underset{}{\sim}x_1'$, $x_{1t}\underset{}{\sim}x_3'$ follows $x_{2t}\underset{}{\sim}x_3'$, further that from $x_{2t}\underset{}{\sim}x_3'$, $x_{2t}\underset{}{\sim}x_2'$, $x_{3t}\underset{}{\sim}x_2'$ follows $x_{3t}\underset{}{\sim}x_3'$, which contradicts the assumption. For the second case that $x_{3t}\underset{}{\sim}x_3'$ holds and $x_{1t}\underset{}{\sim}x_3'$ is wrong we may derive a contradiction in a similar way.

We have now to say something about the invariance transformations. Let H and H' be the groups of homomorphisms of M_{pot} and M'_{pot}:

D2a $t \in H \rightarrow (t: M_{pot} \longrightarrow M_{pot})$ and

D2b $t' \in H' \rightarrow (t': M'_{pot} \longrightarrow M'_{pot})$,

$t \in H$ and $t' \in H'$ are bijections.

The subgroups G_T and $G_{T'}$ of H and H' containing those transformations which leave the theory $T = \langle M_{pot}, M, I_R, I_S \rangle$ and $T' = \langle M'_{pot}, M', I'_R, I'_S \rangle$ invariant satisfy the relations:

D3a $t \in H \rightarrow (t \in G_T \leftrightarrow \bigwedge x \bigwedge y (x \in M \wedge y \in I_R \leftrightarrow tx \in M \wedge ty \in I_R))$

D3b $t' \in H' \rightarrow (t' \in G_{T'} \leftrightarrow \bigwedge x' \bigwedge y' (x' \in M' \wedge y' \in I'_R \leftrightarrow t'x' \in M' \wedge t'y' \in I'_R))$

D3c $e \in G_T$ and $e' \in G_{T'}$ are the unit elements of the groups G_T and $G_{T'}$.

Now we are ready to formulate an important condition which connects the invariance groups G_T and $G_{T'}$ with the relations of empirical equivalence $\underset{e}{\sim}$ and $\underset{e'}{\sim}$ and which has to be satisfied by any empirically relevant group of invariance transformations.

A4a $t \in G_T \wedge x \underset{e}{\sim} y \rightarrow t x \underset{e}{\sim} ty$

A4b $t' \in G_{T'} \wedge x' \underset{e'}{\sim} y' \rightarrow t'x' \underset{e'}{\sim} t'y'$

This axiom say that the relation of empirical equivalence
has to be invariant under the invariance group of the physical theory. Two possible models which are empirically equivalent are always transformed into empirically equivalent possible models. This condition is far from being trivial but
very plausible. We can show that it has a strong intuitive
appeal. Assuming for example that spatial translations are
invariance transformations of natural laws, it seems to be
absurd to assume that two empirically different descriptions
of physical processes become empirically equivalent for the
descriptions of these processes if they are dislocated to a
different place. If two processes are physically different
at one place they cannot be physically identical if they are
brought to another place by the same spatial transformation.
And if we consider other invariance transformations the situation seems to be not much different. Intuitive considerations of such a kind are indeed no sufficient account. We
can only hope that one will be found in future. For the moment we shall just assume that axiom A4 holds.

The relations $\underset{e}{\sim}$ and $\underset{e'}{\sim}$ which hold between empirically
equivalent possible models $x \in M_{pot}$ or $x' \in M'_{pot}$ enable us, to
construct equivalence classes $\bar{x}$ or $\bar{x}'$ of such possible
models, which obtain empirically equivalent descriptions of
the same physical processes.
These are defined as follows

D4 $x \in M_{pot} \quad \rightarrow \quad \bar{x} = \{y \mid y \underset{e}{\sim} x\}$

          $x' \in M'_{pot} \quad \rightarrow \quad \bar{x}' = \{y' \mid y' \underset{e'}{\sim} x'\}$

We shall consider the stroke "—" as an operator defined for
all echelon sets over the two basic sets M_{pot} and M'_{pot}, and
shall define it by an inductive definition which is obtained,
if D4 is supplemented by D5 and D6.

D5   Let B be any echelon set over M_{pot} and M'_{pot} .
Then we have
$$\bar{B} = \{y \mid \bigvee x (x \in B \wedge y = \bar{x})\}$$

D6   Let A and B be any echelon set over the basic sets
          M_{pot} and M'_{pot} .
Then
$$\overline{\langle A,B \rangle} = \langle \bar{A}, \bar{B} \rangle$$

The thus defined operator "—" projects echelon sets over
M_{pot} and M'_{pot} into the corresponding echelon sets over
$\bar{M}_{pot}$ and $\bar{M}'_{pot}$, while the sets $\bar{M}_{pot}$ and $\bar{M}'_{pot}$ consist of classes of empirically equivalent elements of M_{pot} and M'_{pot} ,

i.e. of empirically equivalent possible models.
M_{pot} and M'_{pot} may be considered as sets of physical processes
since empirically equivalent descriptions may be thought of
as describing the same physical process. Thus, instead of
talking about representations of physical processes, we are
rather talking about themselves, and the operator "——"
may be considered as projecting descriptions of physical
entities onto these entities themselves. If this in an
adequate account, there must be a one-to-one correspondence
between the element of M_{pot} and $\overline{M}'_{pot}$ since the identity
criteria of physical processes must be independent of the
description. And indeed we may prove the theorem which says
that there is a bijection τ_t which maps equivalence classes
of T-descriptions of processes into T'-descriptions of the
same processes, or in different terms that there is a
bijection τ_t from $\overline{M}_{pot}$ onto $\overline{M}'_{pot}$. This theorem seems to
be trivial in the light of our interpretation of the symbolism.
Nevertheless it has to be proved in the formal theory. In
order to facilitate this proof we first need a lemma, which
states that τ_t connects empirically equivalent models always
with equivalent models:

T4: $x \underset{t}{\sim} x' \wedge y \underset{t}{\sim} y' \longrightarrow (x \underset{e}{\sim} y \longleftrightarrow x' \underset{e}{\sim} y')$

Proof: Let us assume that $x \underset{t}{\sim} x'$, $y \underset{t}{\sim} y'$ and $x \underset{e}{\sim} y$. The definition D1a of $\sim$ then implies that there is a z' such that
$x \underset{t}{\sim} z'$ and $y \underset{t}{\sim} z'$. If we apply axiom A2 to $y \underset{t}{\sim} y'$, $y \underset{t}{\sim} z'$ and $x \underset{t}{\sim} z'$
we obtain $x \underset{t}{\sim} y'$. Therefore there is a z (namely x), such
that $z \underset{t}{\sim} x'$ and $z \underset{e}{\sim} y'$, which is according to the definition
of $\underset{e'}{\sim}$ (D1b) equivalent to $x' \underset{e'}{\sim} y'$. In the same way as we
have just done here we may show from $x \underset{t}{\sim} x'$, $y \underset{t}{\sim} y'$ and
$x' \underset{e'}{\sim} y'$ that $x \underset{e}{\sim} y$ and have thus proved the lemma.
   We are now prepared to show

T5: τ_t is a bijection from $\overline{M}_{pot}$ into $\overline{M}^r_{pot}$.

Proof: From D4 we may derive that $x \underset{e}{\sim} y$ is equivalent to
$\overline{x} = \overline{y}$ and $x' \underset{e'}{\sim} y'$ is equivalent to $\overline{x}' = \overline{y}'$, such that T4
may be reformulated as: $x \underset{t}{\sim} x' \wedge y \underset{t}{\sim} y' \rightarrow (\overline{x} = \overline{y} \leftrightarrow \overline{x}' = \overline{y}')$.
The expression $X \overline{\tau}_t X'$ however means by definition (D4,D5,D6)
that $\bigvee x \bigvee x' (X = \overline{x} \wedge X' = \overline{x}' \wedge x \underset{t}{\sim} x')$. If we are aware of
this fact, we see immediately from the reformulation of
lemma T4, that $\overline{\tau}_t$ must be a bijection, since if X is an
equivalence class containing x and y, also x' and y' for
which $x \underset{t}{\sim} x'$ and $y \underset{t}{\sim} y'$ hold will belong to a common equivalence

class X'.

By the proof at theorem T5 we have come one step nearer to our goal, to formulate a criterion for the distinction of empirically relevant invariance transformations from empirically irrelevant so called gauge transformations. We have now to ask, if there are groups $\overline{G}_T$ and $\overline{G}_{T'}$ analogous to G_T and $G_{T'}$ which contain invariance transformations of $\overline{M}, \overline{I}_R$ and $\overline{M}', \overline{I}'_R$ respectively.

For this purpose we have to show that $\overline{G}_T$ and $\overline{G}_{T'}$ are groups and particularly that these sets are (not necessarily faithful) realizations of G_T and $G_{T'}$. Later we shall see that these realizations are isomorphic. In order to prove this we start by showing that for $t \in G_T$ and $t' \in G_{T'}$ the relations $\overline{t}$ and $\overline{t}'$ are bijections of $\overline{M}_{pot}$ and $\overline{M}'_{pot}$ respectively.

T6a $t \in G_T$ $\rightarrow$ $\overline{t}$ is a bijection
T6b $t' \in G_{T'}$ $\rightarrow$ $\overline{t}'$ is a bijection

Proof of T6a: Let us assume that $t \in G_T$.
If $\overline{t}$ is no bijection, there will be two pairs $\langle x_1, y_1 \rangle \in t$ and $\langle x_2, y_2 \rangle \in t$ such that according to D4-6 $\langle \overline{x}_1, \overline{y}_1 \rangle \in \overline{t}$ and $\langle \overline{x}_2, \overline{y}_2 \rangle \in \overline{t}$ and that either $\overline{y}_1 = \overline{y}_2$ and $\overline{x}_1 \neq \overline{x}_2$ or $\overline{y}_1 \neq \overline{y}_2$ and $\overline{x}_1 = \overline{x}_2$. Both cases however are not compatible with axiom A4a, the invariance of $\sim$ under $t \in G_{T'}$, since in the first case $\overline{y}_1 = \overline{y}_2$ and $\overline{x}_1 \neq \overline{x}_2$ is equivalent to $y_1 \underset{e}{\sim} y_2$ and $\neg (x_1 \underset{e}{\sim} x_2)$ which plainly contradicts the invariance of $\sim$, and in the second case the situation is completely the same. The proof of T6b is analogous.

If we have seen that the relations $\overline{t}$ for $t \in G_T$ and $\overline{t}'$ for $t' \in G_{T'}$ are bijections, we still do not know that they form groups $\overline{G}_T$ and $\overline{G}_{T'}$. For this purpose we have also to show that for $t_1, t_2 \in G_T$ we have $\overline{t_1 \cdot t_2} = \overline{t}_1 \cdot \overline{t}_2$ and that a similar relation holds for $t'_1, t'_2 \in G_{T'}$.

T7a $t_1 \in G_T \wedge t_2 \in G_T$ $\rightarrow$ $\overline{t_1 \cdot t_2} = \overline{t}_1 \cdot \overline{t}_2$

T7b $t'_1 \in G_{T'} \wedge t'_2 \in G_{T'}$ $\rightarrow$ $\overline{t'_1 \cdot t'_2} = \overline{t}'_1 \cdot \overline{t}'_2$

Proof of T7a: Let us assume that $\langle X, Z \rangle \in \overline{t}_1 \cdot \overline{t}_2$ or what is the same that there is a Y such that $\langle X, Y \rangle \in \overline{t}_1$ and $\langle Y, Z \rangle \in \overline{t}_2$. Then there are $x \in X$, $y \in Y$ and $z \in Z$ such that $\langle x, y \rangle \in t_1$ and $\langle y, z \rangle \in t_2$, and therefore $\langle x, z \rangle \in t_1 \cdot t_2$. From this follows that $\langle \overline{x}, \overline{z} \rangle \in \overline{t_1 \cdot t_2}$ (according to D4-6) or since $\overline{x} = X$ and $\overline{z} = Z$ that $\langle X, Z \rangle \in \overline{t_1 \cdot t_2}$. We have thus shown that from $\langle X, Z \rangle \in \overline{t}_1 \cdot \overline{t}_2$ follows $\langle X, Z \rangle \in \overline{t_1 \cdot t_2}$

Now we know already from T6a that $\overline{t_1 \cdot t_2}$ is a bijection.
Therefore there can be only one X for a given Z and only one
Z for a given X such that $\langle X, Z \rangle \in \overline{t_1 \cdot t_2}$, and if
$\langle X, Z \rangle \in \overline{t_1 \cdot t_2}$ also $\langle X, Z \rangle \in \overline{t_1} \cdot \overline{t_2}$ has to be the case.
    From the two theorems T6 and T7 follows that
T8 $\overline{G}_T$ and $\overline{G}_{T'}$ are groups.
Proof (for G_T): $\overline{G}_T$ is the set of all $\overline{t}$ for which $t \in G_T$.
We know that all $\overline{t} \in \overline{G}_T$ are bijections and that
$\overline{t}_1 \cdot \overline{t}_2 \in \overline{G}_T$ if $\overline{t}_1 \in \overline{G}_T$ and $\overline{t}_2 \in \overline{G}_T$, since
$\overline{t}_1 \cdot \overline{t}_2 = \overline{t_1 \cdot t_2} \in \overline{G}_T$.

A set of bijections however, the product of which belongs to
the same set is in any case a group. If the bijection $\overline{\tau}_t$
maps equivalence classes of possible models $X \in \overline{M}_{pot}$ of one
linguistic representation onto equivalence classes of
possible models $X' \in \overline{M}'_{pot}$ of another representation it will
establish also a one-to-one correspondence of equivalence
classes of models $X \in \overline{M}$ to $X' \in \overline{M}'$. For models are descriptions of physical processes which are compatible with the
theory, and if both representation T and T' are empirically
equivalent, they must admit the same physical process as
physically possible. Since $X \overline{\tau}_t X'$ indicates that the equivalence classes X and X' refer to the same physical process,
it seems to be clear that for any $X \in \overline{M}$ also $X' \in \overline{M}'$, provided
$X \overline{\tau}_t X'$ holds between X and X'. Furthermore we should expect
a similar result for $\overline{I}_R$ and $\overline{I}'_R$: $X \in \overline{I}_R \wedge X \overline{\tau}_t X' \to X' \in \overline{I}'_R$.
We have therefore to prove the following theorems:
T9a $X \in \overline{M}_{pot} \wedge X' \in \overline{M}'_{pot} \wedge X \overline{\tau}_t X' \to (X \in \overline{M} \leftrightarrow X' \in \overline{M}')$

T9b $X \in \overline{M}_{pot} \wedge X' \in \overline{M}'_{pot} \wedge X \overline{\tau}_t X' \to (X \in \overline{I}_R \leftrightarrow X' \in \overline{I}'_R)$

Proof of 9a: Let us assume that
(9.1) $X \in \overline{M} \wedge X \overline{\tau}_t X'$
s has already been noticed in the proof of T5, $X \overline{\tau}_t X'$ is
equivalent to
(9.2) $\bigvee x \bigvee x' (X = \overline{x} \wedge X' = \overline{x}' \wedge x \underset{t}{\sim} x')$
We replace now (9.2) by a logically stronger version which
is implied by $X \overline{\tau}_t X'$ and D1a, T3, and D4:

(9.3) $\bigwedge x \bigvee x' (X = \overline{x} \longrightarrow X' = \overline{x}' \wedge x \underset{t}{\sim} x')$
Should (9.3) not be the case, there would be a $y \in X$ such
that $\overline{y} = X$ and $\neg (y \underset{t}{\sim} x')$ while $x \underset{t}{\sim} x'$ holds, in contradiction to the definition of $\underset{e}{\sim}$ (D1a). (We remember that if
x and y belong to the same equivalence class X, also $x \underset{e}{\sim} y$
has to be the case.)

From (9.1) we conclude that $X \in \overline{M}$, which is by definition
(D4,D5) equivalent to
(9.4) $\bigvee x(\overline{x} = X \wedge x \in M)$.
Axiom A3a yields
(9.5) $x \underset{t}{\sim} x' \rightarrow (x \in M \rightarrow x' \in M')$
and finally we may conclude from (9.3), (9.4) and (9.5) by
purely logical transformations that
(9.6) $\bigvee x' (\overline{x}' = X' \wedge x' \in M')$,
which is by definition (D4,D5) equivalent to (9.7) $X' \in \overline{M}'$.

If we remember that we have derived (9.7) from (9.1),
and that in a similar way $X \in \overline{M}$ may be derived from $X \overline{\sim}_t X'$
and $X' \in M'$, we see that the theorems T9a has been demon-
strated. The proof of T9b is analogous.

We have now to prove just one additional theorem before
we come to the essential conclusions of this section. We
know already that $\overline{G}_T$ is a group but have not yet shown that
it is the invariance groups for the theory $<M_{pot}, M, I_R, I_s>$.
T10a: $\overline{G}_T$ is the invariance group of
$$<M_{pot}, M, I_R, I_2>$$

The proof of T10a is similar to that of T9a, and is not here
given in detail. We have to show first analogoue of
(9.3):
(10.1) $Y = tX \longrightarrow \bigwedge x \bigvee y(X = \overline{x} \wedge Y = \overline{y} \wedge y = tx)$

If (10.1) would not be the case though $Y = t X$ is pre-
supposed there would be an x and an y such that $\overline{x} = \overline{y}$
and $\overline{tx} \neq \overline{ty}$, which is incompatible with axiom A4a.

Using (10.1) instead of (9.3) we may proceed in the
same way as in the proof of T9a.
T9b: $\overline{G}_{T'}$ is the invariance group of $<\overline{M}'_{pot}, \overline{M}', \overline{I_R'}, \overline{I_s'}>$

T9b is obtained from T9a by simply replacing all unprimed
symbols by the primed ones.

We have now to show that the structure of $\overline{G}_T$ is language
independent, or that $\overline{G}_T$ and $\overline{G}_{T'}$ are isomorphic.
T11 $\overline{G}_T$ and $\overline{G}_{T'}$ are isomorphic.
The proof is completely analogous to the proof of T1 and
T2. Theorems T9a, T9b and T10 express the conditions under
which T2 holds.
T11   is the important result of this section. Even if there
is no one-to-one correspondence between the possible models
of a theory in two different linguistic formulations, there
is a group the structure of which can be considered as a
common language independent feature of both representations
of the theory. This invariance group structure cannot be

changed by any reformulation of the theory and is therefore
immune against all conventionalistic strategies. We shall
discuss the meaning and the consequences of this result in
the next section.

In the present section however we have still to answer
one question, which has been left. We have seen the example
of the vector-potential formulation of electrodynamics. There
appeared a kind of transformations which are usually called
gauge transformations. Now we ask: can we generalize this
concept and say in general what gauge transformations are?
We have for this purpose to study the relation between G_T
and $\overline{G}_T$. We may consider $\overline{G}_T$ as a realization of the group
G_T , but not necessarily as a faithful one. It may happen
that different elements of G_T are projected into one
element of $\overline{G}_T$. There may be $t_1 \in G$ and $t_2 \in G$ with
$t_1 \neq t_2$ while for $\overline{t}_1 \in \overline{G}$ and $\overline{t}_2 \in \overline{G}$ we have $\overline{t}_1 = \overline{t}_2$.
This means that $\overline{t}_1 \overline{t}_2^{-1} = \overline{e}$ while $t_1 t_2^{-1} \neq e$.
Thus there are possibly elements different from e which
are projected into $\overline{e}$. These elements form a group by
themselves as is easily seen, since $\overline{e} \cdot \overline{e} = \overline{e}$ and $\overline{e}^{-1} = \overline{e}$.
We want to call this group E_T .

D7 $E_T = \{t \mid t \in G_T \wedge \overline{t} = \overline{e}\}$

T12 E_T is a group.

The proof of T12 has already been given in the text. The
transformations of this group map the element of $M_{not.}$ on
themselves while those of M_{pot} are pictured onto different
ones. This means that the transformations $t \in E$ map the equi-
valence sets $X \subseteq M_{pot}$ with $X = \overline{x}$ for an $x \in M_{pot}$ on themselves, they are intrinsic transformations of those sets.
The relation $\underset{e}{\sim}$ is, as we know, the relation of empirical
equivalence between possible models. Thus $t \in E$ maps possible models on empirically equivalent ones, and that is
just what gauge transformations do. Therefore we can identify the group E with the <u>gauge group</u> of the theory.

5. INVARIANCE TRANSFORMATIONS AND GEOMETRIES

In the preceding section we have shown that empirically significant invariances of physical laws are description independent or language independent. In section 2 we had posed
the question if it is possible to find out the geometry of
space without reference to the language used for the

formulation of particular physical laws. It has been Duhems
and Quines claim, that this is not possible. For Quine there
is no sense in talking about geometry alone; only a conjunction of geometrical and physical laws may be corroborated or
falsified. If we want to attack this "holistic" thesis, we
have to find a connexion between the physical geometry of the
world and its physical invariance properties.

Most physicists would be ready to admit that there is
such a connexion, most philosophers of sciences would deny
that invariances of physics have anything to do with geometry.
For physicists it is quite common to say that space is homogeneous if the physical laws are invariant under the groups
of translations and to say that space is isotropic if the
physical laws are invariant under the group of rotations.
Philosophers of science tend to insists on the conventionali-
ty of spatial metric and to point out, that there is no
apriori reason why one should not introduce any arbitrary
function $d(x,y)$ which satisfies the axioms of metrical geometry (particularly the triangle inequality) in a given
coordinate system.

Physicists however would never do that; they would
only discuss distance functions $d(x,y)$ of a certain kind.
They would for instance never introduce a metrical function
which has less symmetries than the laws of physics have
themselves, but they have never explicitly formulated the
criteria, which they imply intuitively. The philosopher of
science who has not the physical intuition however will
only understand the physicist if these criteria have been
formulated explicitly, what we shall do now.

For this purpose we have first to introduce the distinction between universal and individual predicates, which has
first been made in a similar way by C.R. Popper (1959, §14;
1967 p.35-39). Examples for individual predicates are "x
is an Australian","x is a Sunday's child", while predicates
like "x is a piece of copper" are universal predicates in
Poppers sense. Individual predicates are in some way dependent on proper names. "Australian" depends on the name
"Australia". "Sunday's child" is a less trivial example; for
its definition we first need "Sunday" and we have to ask
ourselves what qualifies a day as being a sunday. There is
certainly no measuring instrument for the determination of
the seven days of the week which can be applied by somebody
who has never known the name of the weekday for any day in
the course of time. If he wants to apply a clock, he has
first to regulate it, and in order to do that, he needs

some information about the weekday for any day known to him.
The reason is, that "Sunday" means "there is a natural number
n, such that n · 7+1 days ago God rested after the creation
of the world". So we have defined sunday historically by
recourse on a name or definite description of an individual.
(It does not matter here, if the report of Genesis 2 is
correct or not.)

     Universal predicates however are completely independent
of any individual thing or situation. They may be applied by
somebody who has completely lost his orientation, who does
neither know, where he is nor the time or the direction in
space. In such case he will still be able to say if a flower
is red, but not if is Wednesday or Friday.

     For practical purposes we should demand, that physical
quantities are defined as universal functions and not as individual functions. It must be possible to apply measuring
procedures in a situation independent way. On the other hand
physicists are only interested in situation independent
informations. If they read the data measured in other
institutes they want to compare them with their own, which
becomes much more difficult, if these data are formulated by
individual predicates. We may also express this attitude of
the physicists by saying that physicists use lawlike or
nomological predicates in their physical language and that
only universal predicates are nomological. The reader, if
he is a physicist, may ask here: are there not vectors
and tensors in physics? And are there not fields? Neither
a vector like force $\vec{f}$ nor a field - even if it is a
scalar field - are invariant scalars, an electrostatical
potential has not to be isotropic or homogeneous. The answer
may be the following. First, instead of using vectors like
$\vec{f}$ we may equally well apply functions like $f(x)=(\vec{f},\vec{x})$,
the scalar product of the vector with the position vector,
and similar considerations are possible in the case of
tensors. Second, in the case of fields we have to introduce a variable for the physical system as an additional
argument of the function - say $x \in M$, the variable for the
possible model in section 3 an 4. In another paper I have
discussed the transformations of vectors, tensors and field
functions in detail (A. Kamlah 1979, p.105-113), and this
extensive discussion cannot be included in the present paper.
For the moment, therefore, I cannot do more than simply
to inform the reader about the fact that all physical
functions which are in actual use in non-relativistic or
special relativistic physics can - if all variables are used,

on which the values of these functions are really depended
- be interpreted as scalar functions which have the same
invariance properties as physics in the non relativistic
or special relativistic frame.

We may now identify universal predicates with those
predicates which are themselves invariant under the invariance
transformations of the natural laws. This seems to be the
essence of universality. We use universal predicates since
we are interested in natural laws. Any statement which contains only universal predicates as logical constants will
automatically have all invariance properties of the universal predicates themselves and is therefore a good candidate
for a natural law, while a statement which violates the
known invariances, surely will be no law of physics. In the
special case of Galilean- or Lorentz- invariant physics
"invariant" means "independent of the inertial system"
or independent of position, direction and velocity, we
could also say "independent of the situation". We are eager
to learn from nature, what is common to similar situations.
We want to put down our knowledge in a language which
enables us to apply it in any given situation. Therefore
we need a situation independent language.

We are prepared to formulate a criterion (Criterion
of invariant description): <u>Apply only such description of
nature, which contain only predicates which are invariant
under the invariance transformations of nature.</u>

These consideration and particularly the formulated
criterion has now to be applied to the metric of space,
time and space-time. It is easy to see that many alternative
metrizations are eliminated by the criterion. Take for
example following metrization on the surface of the earth

$$ds = dl \; \frac{1}{\cos^2 \left((90^\circ - \phi)/2\right)}$$

where dl is the usual and ds the alternative length of a
very (infinitesimal) small distance. The angle ϕ is the
geographical latitude. This metrization would have the
advantage for the surveyor to lead to a flat representation
of the earth' surface, known as stereographic projection.
It would however violate the criterion of invariant description, since in it the spatial distance function is not invariant, while it is one in another description - namely in
the usual description. It would also be difficult to apply
the new metric. If on board of a ship somebody want to know
if a cupboard would fit into a cabin he could only solve

this problem by applying the new metric ds if he knew
the exact position of the ship on sea. Otherwise it would
be impossible for him to determine the length, depth and
width of the cupboard, the cabin and the measures of its
door. If he has answered the question in the harbour, the
position of which is well known, and if he has actually
brought in the cupboard into the cabin, he still does not
know, without complicated calculation, if he can conclude
from his measurements at another position on sea, that it
will be possible to bring the cupboard out again through
the door. All theses problems become very trivial however,
when the usual spatial distance function is used. Applying
it we get situation independ answers to situation independent questions.

Another famous example, which is taken from time
measurement, is the proposal to measure time by the pulse
of the Dalai Lama (M. Schlick 1925, §11). Using this holy
"standard clock" we would obtain a time distance function
$d(t_1,t_2)$ which is not time-translational invariant. The
example is discussed in more detail in A. Kamlah 1973,
p. 241-245.)

Thus we see that physicists have good reasons only to
apply invariant spatial and temporal distance functions and
in the case of special relativity, where no such invariant
exist, an invariant space-time distance function. In any
case the space-time distance function should have all invariance properties of the natural laws. If this maxim is
obeyed, the freedom of choice for possible metrical functions
is narrowed down considerably, and what is important for us
now: physical geometry will share all invariance properties
with the natural laws.

We now remind ourselves of the original question posed
in section 2. There we asked how it might be possible to
falsify the geometry of physical space without recourse to
the particular physical laws and the language used for
them. In section 3 and 4 we have already seen that the
symmetry of natural laws, their invariance properties are
independent of the chosen physical language. In the present
section we have now found a link between the symmetric of
nature and those of geometry by the criterion of invariant
description which applies particularly to the spatial
distance function. If this criterion is accepted, we come
to the conclusion that also the invariance properties of
physical geometry are independent of the language of physics.
We have thus partially to deny the Duhem-Quine-thesis.

The falsification of physical geometry – as far as its
invariance properties are concerned – is indeed independent
of the falsification of particular physical laws.

Finally one may add the question, how far geometry is
determined by its invariance properties. In the case of
absolute geometry (euclidian, elliptic, spherical or
hyperbolic geometry) this determination is complete if the
topology is given. In other cases where we deal with less
symmetrial metrical fields, we still have many possible
choices for the distance function, even if the invariances
are known. The metrical field may in all properties which
are not determined by the symmetry of the natural laws
depend on the language chosen for physics just as has been
claimed by Duhem and Quine and particularly by H. Putnam
(1963, p. 242-255).

The symmetry however which is common to all admissible
distance functions is the invariance of natural laws. In any
metrical and continuous geometry the metrical field will
have at least the same spatial invariances as the complete
physical theory, if only invariant distance functions are
admitted. The thus defined geometry may however be more
invariant than nature, for instance, if some forces
violating an invariance obeyed by all other forces do not
influence the measuring rods. The class however of all
admissible distance functions has just the spatial invariance of the whole nature. Thus the group of all spatial
transformations which leave invariant the natural laws
characterizes the geomteric properties of the world in a
unique and unambiguous way as an invariance group for all
possible metrical fields. By no conventionalist reformulation of the physical theories we then introduce a
geometry which has not the spatial invariances of nature.

So conventionalism can only be upheld in a more modest
form. If physical concepts of any kind are admitted, any
geometry may surely be obtained by a suitable distance
function. This statement is the thesis of what Grünbaum
calls "trivial semantical conventionalism" (A. Grünbaum
1963, p.27). But there is also a stronger thesis which
is false, as we hope to have shown in this paper. It
claims: We cannot specify the kind of distance function to
be admitted in such a way that in some possible worlds only
some geometries become possible. And above all our specification is not arbitrary but expresses in the Euclidian
case (and in similar cases, as may easily be shown) a
feature of measurement, namely reproducibility, that means

the independence of the method from the knowledge of ones
own position in space.

REFERENCES

Balzer, W. and A. Kamlah: 1980, 'Geometry by Ropes and Rods',
  Erkenntnis 15, 245-267.
Einstein, A.: 1949, 'Reply to Critisicm' in P.A. Schilpp (ed.),
  Albert Einstein: Philosopher-Scientist, The Library of
  Living Philosophers, Evanston.
Grünbaum, A.: 1963, Philosophical Problems of Space and Time,
  A.A. Knopf, New York.
Kamlah, A.: 1973, 'Invarianzgesetze und Zeitmessung'
  Zeitschrift für allgemeine Wissenschaftstheorie, 4, 224-260.
Kamlah, A.: 1976, 'An Improved Definition of 'Theoretical in
  a Given Theory', Erkenntnis 10, 349-359.
Kamlah, A.: 1979, 'Zur Eindeutigkeit von Zeit- und Längen-
  messungen', in W. Diederich (ed.), Zur Begründung physi-
  kalischer Geo- und Chronometrien, Universität Bielefeld,
  Schwerpunkt Mathematisierung, Materialien 13.
Poincaré, H.: 1952, Science and Hypothesis, Dover Publ.Inc.
  New York.
Popper, C.R.: 1959, The Logic of Scientific Discovery,
  Hutchinson & Co., London.
Popper, C.R.: 1967, Logik der Forschung (2nd ed.), Mohr,
  Tübingen.
Putnam, H.: 1963, 'An Examination of Grünbaum's Philosophy
  of Geometry' in B. Baumrin (ed.), Philosophy of Science
  (The Delaware Seminar, Vol. 2), Interscience Publishers,
  New York, pp. 205-255.

Wolfgang Deppert

OUTLINE OF A THEORY OF SYSTEM-TIMES [+]

INTRODUCTORY REMARKS

Since Kant, at least, widespread acceptance has been gained by the basic philosophical view that space and time should no longer be understood as characteristics of the world as such but rather as conditions set upon the possibility of human experience. The appearance of non-Euclidean theories underlified the fact that other and further conceptions of space were competing with the Euclidean geometry which Kant held to be the sole viable geometry, and there began to be talk in terms of concepts of space. When Einstein then, in his General Theory of Relativity, fused space and time into a four-dimensional manifold based upon Riemannian geometry, it became possible to talk also in terms of concepts of time — or more exactly, in terms of spacetime concepts.

When I talk in this paper in terms of the concept of time I am not taking the above-mentioned connection as my point of departure. Strictly speaking, my procedure is in fact quite the opposite; that is, instead of referring to a concept of space, I attempt first of all to set up a concept of time, though I am aware in doing so that this concept of time implies particular conceptions of space — conceptions which I cannot, however, go into here.

The one-dimensionality of the time-continuum is nowadays generally accepted, and because of this it is impossible to argue analagously from the various non-Euclidean concepts of space to various concepts of time. This is only possible within the framework of metrization procedures of the concept of time, and this is the path which the present paper sets out to plot. In doing so, I assume a comparative — or, as it is called topological concept of time, as is the case in all metrization procedures (c.f., CARNAP (1966), REICHENBACH (1928), STEGMÜLLER (1970)). This comparative concept of time contains within it notions

D. Mayr and G. Süssmann (eds.), Space, Time, and Mechanics, 195–224.
Copyright © 1983 by D. Reidel Publishing Company.

of simultaneity, and of 'earlier than' or 'later than'. The
concept of time presented here arises from the ambiguity
of the metrization procedure with an unchanged topology
of time.

1. THE METRIZATION OF TIME

A metrization is a mapping, called d, of a domain D consisting of objects p_i into an ordered set of numbers,
such that for each object p_i there exists a unique element
$d(p_i)$ of the chosen set.[1] The set may be the set of natural
numbers $\mathbb{N}$, the set of integers $\mathbb{Z}$, the set of real numbers
$\mathbb{R}$, or the n-tuple of these sets of numbers, as in $\mathbb{R}^n$.
The mapping used for metrization is also called 'functor'.

The first problem to be solved is the determination
of the objects p_i and the choice of their domain D. The
objects of metrization of time can be either events or
processes. Since it is not practical to metrize events as
the first step, I am restricting domain D to processes only. If I then choose as my set the positive real numbers,
I have

$$d(p_i) = 0$$

for all processes p_i. This follows from the notion that the
duration of a process can be stated with the help of positive numbers.

To specify the function d we follow the standard time
metrization procedure (TMP) in three steps (c.f., CARNAP
(1966), REICHENBACH (1928), STEGMÜLLER (1970)).

1. Choose a process p_o which is assigned the number
one

$$d(p_o) = 1 \qquad \text{(unity statement)}.$$

2. Find a rule for deciding which of the processes p_i
and p_j should be mapped onto the same number, so that

$$d(p_i) = d(p_j) \text{ (equality statement)}.$$

3. Look for a combination rule for processes, so that the following condition (which will be called the additive extensiveness condition) holds:

$$d(p_i \circ p_j) = d(p_i) + d(p_j),$$

where the sign ' o' designates the special combination procedure which combines the process p_i into the new process $p_i \circ p_j$ (scaling statement).

The necessity to make statements on the unit of measure, on the equality of objects, and on the scaling of the measure arises from formal reasons relating to general metrization procedure. The special form and content of these statements do not depend only on empirical facts, for there is great freedom of choice in carrying out the individual metrization steps. This is especially true with regard to the TMP, so that there are in fact several possibilities for time metrization.

This is not the place to discuss the arbitrariness of equality and scaling statements. I shall restrict myself instead to a consideration of the unity statement, and from this it will be seen what far-reaching conclusions an arbitrariness of statements gives rise to.

The time-measure should be introduced here by means of a fundamental metrization, i.e., it should not be derived from another measure, such, for example, as the metrization of length. In this case, the process p_o, for which $d(p_o) = 1$ holds, must be a periodic process.[2] I take a periodic process to mean a process in which certain combinations of features occur repeatedly, e.g. the building of nests by birds in Spring, a certain arrangement of sounds in a piece of music, a certain water-level in the river Elbe, or a certain distribution of pressure in the cylinder of a combustion engine. The question now is: which process do we choose from the wealth of periodic processes for the unity statement?

The obvious thing to do is to look for a process with periods of exactly the same duration. But how are we to determine that a process has this quality? To do so we would need to employ another process whose periods we already know to be exactly of the same duration, and this entails us in an infinite regression.

To avoid this dilemma, we introduce the notion of
periodic equivalence proposed by Carnap[3]. Two processes are
said to be periodically equivalent if the ratio of their
frequencies remains, within a given degree of accuracy, the
same.[4] The notion of periodic equivalence meets the conditions of an equivalence relation, i.e., of a class-producing
relation. As a result of this the set of all periodic processes is divided into empty-intersection classes of periodically equivalent processes (PEP).

The question as to which periodic process belongs to
which periodic equivalence class has to be decided by empirical investigation, although even here the finite degree of accuracy of measurement makes it impossible to proceed without arbitrary statements.

For a special given accuracy of measurement, and
taking into account systematic correction-factors, it is
an empirical fact that oscillations of atoms, molecules,
and crystals belong to the same class as the oscillation of
a pendulum and the revolution of the earth on its axis or
its orbiting around the sun. This is only true, however,
when the conditions under which these processes take place
remain the same. The class of periodically equivalent processes (PEP-class) which contains the processes quoted is
called here the physical class. Since as yet it has not been
possible to find discrepancies in periodic equivalence between gravitational clocks and electro-magnetic clocks, I
am assuming that it makes sense to talk in terms of one
physical class[5].

Now there are certainly a great many periodic processes which have no element in common with the physical
class. This is true, for instance, of my heart-beat, which
is, however, periodically equivalent to the pulse-beat in
my right or left forearm and indeed to a wealth of other
periodic processes in my body. Every organism will possess
at least one class of periodically equivalent processes
which has nothing in common with the physical class. We
shall call these classes organic classes of periodic processes. The set of these organic classes is obviously too
large to be surveyed.

Everywhere periodic processes are to be found we can
look for their equivalence classes, be they periodicities
of biological, psychological, sociological, historical,
economic, technical, chemical, or physical systems. In principle, any representative of any equivalence class is capable of providing the process p_o for our unity statement
$d(p_o) = 1$. When this representative is taken from the physical class, then the time-measure gained from it should be
called <u>physical time</u>, and when on the other hand organic
classes are used, we should talk in terms of <u>organic times</u>.
Similarly, one could talk in terms of psychological, sociological, historical, economic, or other times.

What are the arguments which speak in favour of singl-
ing out physical time as the generally valid measure of
time?
The main arguments for this can be summarised in the following hypotheses:[6]
A 1.   The physical class is by far the largest class of
       periodically equivalent processes, and all other
       classes consist in general of only one element.
A 2.   The laws of nature assume their simplest form if one
       applies the physical class for TMP.

In the next section we shall show that both these
hypotheses fail to convince as arguments for distinguishing
physical time from all other possible times.

2.   THE SET THEORY EXTENSION OF THE CONCEPT OF TIME

The arbitrariness in the TMP of physical time gives rise to
the supposition that the notion of time is a generic one,
i.e., since there are several possibilities for the metrization of time from which we can choose, then there are
also several times. In principle, we could change the concept of time everywhere we find arbitrariness in the TMP.
For the purposes of this paper, it will be sufficient to
discuss the various possibilities in the case of the unity
statement.

To begin with, we shall criticize argument A 1. by
posing the question of the size of the classes of periodic

processes. No one has ever counted the processes which are
periodically equivalent to my heart-beat. Certainly, the
number of bodily processes occurring in rhythm with my
heart-beat is too large to estimate - after all, a change
in blood-pressure has effects which go right down to the
finest capilaries of the blood-vessels. Further, one of the
more recent suppositions of modern science is that there
are no such things as isolated points[7]; this would mean, for
example, that the processes occurring in my body have an
effect on the physical world as a whole, even if such effects are in general of unmeasurable intensity. Thus the
electrons periodically set in motion by my pulse-beat should
according to the Pauli exclusion principle, periodically influence the electrons on the star Sirius, since this exclusion principle states that no two electrons can ever be
in the same state. At any rate, the argument that the physical cláss is larger than the class of processes periodically equivalent to my heart-beat cannot be proved as long
as no notion of the set of all processes, and a counting-
procedure within that set, can be produced.

Since this line of thought can be carried through for
all classes of periodic equivalence, argument A 1. for a
clear distinction of the physical class with regard to
class-size, remains unconvincing. Even if agreement could
be reached as to an order of magnitude of classes, we still
could not accept a quantitative argument such as A 1., since
with regard to the meaning of the concept of time we require a systematic argument for the choice of time-metriz-
ation, i.e., an argument consistent with the concept of
metrization.

Instead of analysing argument A 2. at this stage, it
will be more fruitful to introduce first the notion of
system-space, i.e., the space in which all processes of a
class take place. What is meant by space here is quite generally the manifold of features within which periodicities
are determined. A pair, composed of a class of periodic equivalent processes (PEP) and of its associated system-space,
we shall call a PEP-system.

One particular reason for introducing time metrization is to enable us to give numerical time-values to
non-periodic processes too, i.e., we want to be able to
describe non-periodic processes, too, within the PEP-system.

But which of these processes can be assigned to a PEP-system?
We can find an answer to this question if we extend the
notion of periodic equivalence to include processes which
themselves show no periodicity of their own but which can
be repeated in toto. The extended version of the notion of
periodic equivalence now runs as follows:
Two processes are said to be <u>periodically equivalent</u> when
the number of immediately sequential repetitions of the
one process is always, whenever the experiment is made, the
same – in relation to a chosen degree of accuracy – as a
given number of the other process. It is quite possible
here that one of the two processes will only occur once.
This extended notion of periodic equivalence, also, is a
class-forming equivalence-relation, and everything that has
been said up to now about PEP-classes, system-space, and
PEP-systems should henceforth be understood in terms of
this extended notion of periodic equivalence.

As a result of this notional extension, the PEP-system
defined, for example, by the physical class, can be understood as the physical world, if we take the physical world
to be the set of processes which would, given the same conditions, have the same physical duration.[8] But the PEP-
class of the stroke of a piston-engine defines a special
PEP-system and, similarly, organic PEP-systems are defined
by organic classes.

The foregoing examples raise the crucial question of
the calssification of processes which cannot be repeated.
Clearly, it is assumed here that agreement has been reached
on what it can mean to say of a process that it can be repeated, and given this premise, a non-repeatable process
cannot be assigned so easily to a PEP-system, unless it is
composed of partially repeatable processes. In general,
though, it may be supposed that a non-repeatable process,
i.e., a process which from the beginning cannot be assigned
to any one particular PEP-system, stands as its own PEP-
system. An example of such would be any human being, whose
individual life is certainly not repeatable.

PEP-systems can also be sub-systems of other PEP-
systems. In order to describe the two fundamentally different possibilities, the following terms have been chosen:

2.1.   A PEP-system whose PEP-class is a subclass of an-
       other PEP-class shall be called a PEP-subclass system.

2.2.   A PEP-system whose system-space is a subspace of an-
       other system-space shall be called a PEP-subspace
       system.

A PEP-class whose processes arise from gravitational interaction is a subclass of the physical class. Gravitational
interaction thus allows us to define a PEP-subclass system
of the PEP-system of the physical world. For gravitational
interaction on a large scale, let us call this the cosmological system. As an example of a PEP-subspace system, we
can take a piston-engine, whose system-space is a subspace
of physical space.
Each PEP-class distinguishes a special time metrization provided one chooses a representative of this class for the
unity statement. Thus each PEP-system can clearly be assigned a system-time [9] if one chooses the time metrization
determined by the relevant PEP-class. The system-time of
the physical world, for example is the well-known and universally employed physical time.
Obviously, this notion of time is different from the many
other system-times, such as that, for example, of a steam-
engine, of my own nervous system, or of any other biological
PEP-system. Each of these different system-times can be
correctly defined in the same way as physical time. As far
as the equality statement with regard to physical time is
concerned, this is wholly unproblematic as long as temporal
data are not connected in any way with spatial data. The
problem of simultaneity in different locations only arises
when one talks of time at a certain position, i.e., when
the time needed for the information-transfer enabling temporal comparisons in different places is finite.[10] Nowadays we are convinced that in the physical world the upper
limit for velocity of information propagation is given by
the speed of light in a vacuum (problems arising from the
existence of tachyons will not be entered into here). The
question of whether there exists for all PEP-systems such
a maximum velocity of information-propagation is still entirely open. In order to throw more light on this problem
it is necessary above all to know more about the constitution and metrics of system-spaces. This is an aspect
which I will discuss elsewhere, since here I want to re-

strict myself entirely to the problem of time. With regard
both to the system-times as defined here and to the standard
time metrizations, the scaling statement is assumed to be
additive extensive.

The search for a relationship between different notions
of system-time presents an extremely difficult problem. In
most cases it will be impossible to find a transformation
rule for changing from one system-time to another, as,
for example, from the system-time of an engine to that of
a human being. Only in those cases where a special subsystem
structure exists would it be possible to find certain relations between the subsystem-times of one PEP-system consisting of several subsystems. This will be possible in particular within PEP-subclass systems. Thus, for example, it is
an empirical fact that the times definable by purely electro-magnetic periodic processes and purely gravitational periodic processes can be seen as identical. Grünbaum gives
a rather different, very interesting example,[11] without,
though, expressly referring to the concept of system-time.
In his example he distinguishes between Newtonian time t,
which is intended to represent precisely that time-scale
within which Newtonian laws are valid, and diurnal time T,
which is defined by choosing a mean sidereal day as unit
(for the purposes of the argument it is immaterial whether
a mean solar day or a mean sidereal day is chosen as unit).
Since the revolution of the earth with regard to Newtonian
time is certainly not regular, the functional relation between t and T is surely not stateable in terms of a linear
function, nevertheless it can be assumed that a function

$$T = f(t)$$

can be discovered which describes this relation. If in
one's calculations one takes account of the friction-losses
caused by tidal impulses, then the main cause of the slow-
ing down of the earth's rotation in respect of Newtonian
time t is probably already determined.

With this example we have the case where the class
which we previously described as a physical class breaks
down into many subclasses as soon as greater accuracy is
achieved in those observations by which the PEP-classes are
determined. Thus it is simply a question of convention how
large one chooses to make the upper class which receives

the name of physical class. It is my view, though, that this title belongs best to that class of which one believes that the transformations between the system-times of the relevant PEP-subclass systems can be stated through physical laws. I shall talk later and in more detail of the special relation between lawlikeness and system-time. In Grünbaum's example, the introduction of diurnal time T defines not only a PEP-subclass system of the physical world, but also at the same time a PEP-subspace system, since the processes periodically equivalent to the earth's rotation are limited to the field of space belonging to the earth's sphere of influence. Grünbaum shows that the description of particle-movement in diurnal time T turns out to be more complicated, since in the formula for accelleration

$$\frac{d^2r}{dT^2} = \frac{1}{\left|f'(t)\right|^2} \frac{d^2r}{dt^2} - \frac{f''(t)}{\left|f'(t)\right|^3} \frac{dr}{dt}$$

with r as spacevector, $f'(t) = \frac{dT}{dt}$ and $f''(t) = \frac{d^2T}{dt^2}$

the velocity-dependent term $\frac{-f''(t)}{\left|f'(t)\right|^3} \frac{dr}{dt}$

still appears. And this means that in the equation of motion for a particle of matter in the diurnal time description a further velocity-dependent formula must be taken into account, which is not the case in the purely Newtonian description.

We are now in a position to discuss Carnap's and Stegmüller's argument, which is that
A 2.: Natural laws assume their simplest form if one uses the physical class for time metrization.

From Grünbaum's example of different time metrizations it becomes clear that there are indeed time metrizations through which natural laws in the Newtonian form in fact become more complicated, but these metrizations take place within the physical class, i.e., the rival PEP-classes are in this case subclasses of the physical class. Argument A 2. however, was not intended to meet such cases, for its purpose was to distinguish the physical class from other

PEP-classes, such as the organic classes.

If it is the case, though, that no transformations can be stated between certain different system-times, as must be assumed between physical time and organic times, then nothing very precise can be said about the form of physical laws of nature if one wants to present them in an organic-time; further, it is even questionable whether physical laws of nature can be mathematically formulated at all with regard to suchlike time metrizations, so that the term "law" could no longer reasonably be used with such badly chosen metrizations. And if one cannot speak of laws, then one cannot of course speak of the simpler or more complicated form of such laws. This makes clear that arguments A 1. and A 2. are not sufficient to distinguish time-metrizations with the aid of the physical class from those possibilities of time metrizations feasible by means of other PEP-classes.

The notion of system-times developed here, however, provides a perfectly natural explanation of why it is reasonable to establish a time metrization based on the physical class for the description of the physical world. For we have seen that within the concept of system-times each PEP-system is assigned precisely one system-time, i.e., according to the definitions presented here, the system-time of the physical world is precisely that one which is metrized with the aid of a representative of the physical class.

From the way in which each PEP-system thus receives its own system-time one is almost inclined to believe that with this justification of physical time metrization a liberal principle is proving to be true, i.e., that a procedure is not justified by demolishing or even prohibiting other points of view but by allowing everyone to choose those points of view of their own which they wish to have as criteria. This is illistrated here, where each PEP-system is assigned its own system-time and the laws obtaining within the system become discoverable according to this system-inherent criterion.

3.   CONSEQUENCES FOR THE PHILOSOPHY OF SCIENCE

A fundamental step in the direction towards showing that
time is not an absolute entity (i.e., independent of other
physical entities), was made by Einstein. He showed that
the introduction of a time-scale is only meaningful in relation to a special coordinate system of space.(EINSTEIN
(1969)) We shall now show that Einstein's conception of the
relativity of time fits easily into the concept of system-
time as it is developed here.

If we regard the processes of the physical class with
respect to changing coordinate-systems which have non-
vanishing relative velocities, then the ratio of the frequencies of these processes also changes, in accordance
with Einstein's theory of relativity, with the changing of
the coordinate-system, thus leading to the conclusion that
different periodic processes which are described by means
of different coordinate-systems cannot be counted as belonging to one and the same class because the frequency-
ratios of these processes change if the coordinate-system
chosen to describe a process is changed. This, however,
contradicts our assumption that these processes arise from
the physical class. In order to build up the physical
class, for example by trying to determine by empirical means
which periodic processes are periodically equivalent to
the movement of the earth around the sun, we have therefore
to refer to a chosen <u>fixed</u> coordinate-system, and in the
conceptual framework developed here the concept of the
physical world as PEP-system of the thus discovered physical
class is based upon that chosen fixed coordinate-system.
Indeed, it turns out that the physical class remains stable
under Lorentz-transformations and that the PEP-systems obtained in this way (based on different coordinate-systems)
can be subsumed under the concept of the physical world.

Einstein's General Theory of Relativity states that
all laws of nature - that is, those laws which describe
the physical world - should be stateable independently of
chosen coordinate-systems.[12] In order to clarify how the
principle of relativity fits into the framework of system-
times, it is necessary first to discuss thoroughly the
question of the relation between time and law.

I have already remarked that physical laws with respect to an organic time are not only very complicated but
probably also not stateable at all in mathematically complete form. Further, I have chosen the concept of system
in such a way that a system is constituted essentially by
the system-time belonging to it, since the corresponding PEP-
class by means of which that system-time is first defined
determines the whole system. Besides this, the <u>laws</u> which
obtain in a system characterize this system, since the
properties of a system are determined by its laws. Thus the
laws which are valid for a system characterize in the same
way a system such as system-time. For this reason I wish to
talk in terms of <u>system-laws</u>.

Seen in this light, it seems perfectly natural that
system-laws can only be discovered when the empirical investigations of a PEP-system have as their basis the system-
time belonging to that system. If the laws discovered in
this way are to characterize the whole PEP-system then they
must be able to show to be independent of the choice of coordinates within the system or of any other descriptive base.[13]
If this should not be the case, then the system-laws would
change according to descriptive base, and this would mean
that they would not characterize the <u>whole</u> system and thus
could not in fact be system-laws. This is the generalisation
of Einstein's general principle of relativity with regard
to PEP-systems. Since this generalisation arises when system-
laws are required to characterize in toto the PEP-system to
which they belong, I shall call this generalised general
principle of relativity the <u>totality principle of laws</u>.

The concept of system-law can be formulated more exactly with the aid of this totality principle:
<u>System-laws are propositions which contain as their essen-
tial components[14] system-defining elements</u> (constants or
variables), <u>and which fulfill the totality principle</u>. In the
philosophy of science there is still no formulation of the
concept of a law which corresponds adequately to the standard use of this term in science.[15] It is my impression that
the concept of system-law as defined here <u>does</u> meet these
requirements.

The term law of nature, in current use nowadays, is intimately connected to the notion of a natural lawfullnes
possessing in its essentials the attributes of a rational

God or of absolute truth: omnipresent, immutable, eternal,
and all-encompassing. The explication of a term such as law
of nature runs up against all the problems connected with a
notion of the world as a totality, and it is precisely the
problems of scope which one meets here that have up to now
prevented an acceptable formulation of the concept of lawlikeness discussed by Stegmüller.

In the presentation of the concept here, the notion of
system-laws and thus which produce the totality of the system.
If we wish to examine this reason more closely, we well not be
uted by means of the PEP-classes. This means, though, that
the scope of system-laws is restricted to the systems they
belong to.

A given system-time can, indeed, be seen as the mani-
festation of a relation-producing connection of effects,
which itself is the raison d'etre of the PEP-system, or in
other words: the ontological reason why a system-time can
be introduced is the same as that which enables us to set up
system-laws and thus produce the totality of the system. If
we wish to examine this reason more closely, we will not be
able to avoid metaphysical assumptions. The definition of
the concept of system-law does not, however, require us to
be <u>explicit</u> about such assumptions.

As is the case with the problem of lawlikeness, the
problem of induction of laws by means of the construction
of PEP-systems and their system-laws can be seen in a new
light. For if we first find a PEP-system, then it is clear
that there <u>are</u> system-laws which characterize this PEP-
system. The problem of induction, thus, no longer consists
in the inference of laws from individual measurement-data
– rather, the problem of finding laws can be seen to consist in the following two questions:

1. What kinds of law-structures are stateable which meet
   the invariance requirements arising from the totality-
   principle?
2. How can individual measurement-data be related to such
   kinds of law-structures in order to arrive at system-laws?

Suchlike problems, and the question of whether the concept of system-law as it is worked out here will stand up
to the objections compiled by Stegmüller, will be thorough-

ly discussed in another paper.

4.   SCIENTIFIC CONSEQUECES

The scientific significance of the concept of system-time
is to be found in empirical and conceptual fields.

I have already indicated that the concept of system-
time opens up the possibility of producing a satisfactory
definition for the concept of law, i.e. by means of the
system-law. This means, however, that a wealth of new questions arise with respect to empirical investigations too, and
in just those areas where systems are constituted by means
of PEP-classes. These systems may be biological, psychological, economic, technical, chemical of physical in nature. It
can indeed be expected that new laws can be found in such
systems if the corresponding system-times are introduced.
After all, the use of the system-time "physical time" in
Physics has led to great successes in the discovery of
physical laws. Since the physical class is just one of many
PEP-classes in the outline presented here, it may reasonably be thought that the investigation of other PEP-classes
by using their assigned system-times could lead to similar
successes. The concept of system-time obliges us to make
two distinct empirical steps in our investigation:

4.1.   <u>To search for PEP-systems</u>

4.2.   <u>To search for system-laws within these PEP-systems</u>

The first step can be sub-divided into the following
questions:

4.1.1.   What classes of processes are PEP-classes, and
         which and how many processes belong to them?

4.1.2.   In what space regions do the processes of the PEP-
         classes occur, i.e., which PEP-systems are consti-
         tuted by means of PEP-classes?

4.1.3.   What subsystem-structures exist in PEP-systems?

4.1.4. What relationship exists in the set of all PEP-
systems found in this way, and which transformation
laws can be stated between system-times?

It can be said without further investigation, for example,
that for each organism there is at least one PEP-system. It
is further true that the subsystems of the organism which
are distinguished by a particular cellstructure, i.e., the
organs, are PEP-subspace systems, and it may even be assumed that in the case of organs one is dealing with PEP-
subclass systems. In this regard, for example, F.A. Forsgren
established as early as 1927 that the liver-functions are
periodic. Biological, psychological, and medical investigations have already pointed to the existence of a
wealth of special rhythms within the organism. Ritchie
R. Ward, in his book The Living Clocks (WARD (1971)) has
given a very readable compilation of the most astonishing
discoveries in this field. These discoveries revolve essentially around the so-called circadian rhythms. The term
'circadian', coined by Franz Halberg, is intended to mean
that the duration of a full cycle is approximately 24 hours.
Such circadian rhythms can be found in single-cell organisms, green plants, animals at all stages of development,
and in human beings. These rhythms have been found to be
widely independent of external time-indicators such as
light-darkness cycles. Changes in external metabolism-
conditions, however, give rise to serious changes in rhythm.
These empirical findings indicate that processes within
the organism are responsible for the maintenance of circadian rhythms, and this is why one speaks of biological
clocks.

In most cases the localization of such time-indicators
in the organism gives rise to serious difficulties. Janet
Harker[16], however, has succeeded in showing that in the
lower esophegeal ganglion of cockroaches there are four
neuro-secretory cells which are responsible for giving
these creatures a certain activity-passivity rhythm, which
though normally adapted to the rhythms of light and dark-
ness, even continue for days on end during permanent il-
lumination. By means of a slow phase-displacement of an artificial light-darkness cycle over the day-night cycle, the
circadian rhythm of the cockroach can be altered, and by
transplanting some or all of the four neuro-secretory cells,

Janet Harker was able to show that these cells in the lower esophegeal ganglion are responsible for controlling the activity-passivity cycle. She was also able to show that the rhythmic impulses are carried in the cockroach's blood. Thus with the ciruculation of its blood the cockroach becomes a PEP-system.

In the experiments described by Ward the tacit assumption is always made that there is such a thing as <u>objective</u> (physical) time, and that organisms need internal clocks in order to orient themselves in a world which is controlled by precisely this time. Given this premise, researchers' interests are dictated primarily by the following questions: how can a living organism measure time? Where in the organism is the biological clock located? What serves to adjust the biological clock? In the system-time scheme, however, such questions are formulated in the following manner: what PEP-systems can be found in the organism under investigation, and what PEP-sybsystem structure is present?[17] What causes synchronisation and desynchronisation between system-times within the organism and outside the organism to take place, as in relation to the physical world? What can be said of the genesis of the organism's PEP-systems?

It is my belief that the concept of system-time allows one to be more rigorously systematic in ones investigation of biological rhythms. It seems, for example, obvious to assume that it is the organism's ability to synchronise the system-times of the PEP-systems it contains which makes it a totality. If it were the case that the human being's synchronised organic-time correlated with his awareness of time, this would lend support to Kant's view that the conditions necessary for the constitution of self-awareness would no longer be met if two different intuitions of time occured for one and the same process in a being who could be considered as an organism <u>with</u> self-awareness. This consideration may well suggest that sleep or other unconscious states can be traced back to a desynchronisation of certain PEP-systems within the organism. If the above assumption is true, then the organism as a whole could well be endangered when the synchronisation of its PEP-systems no longer seems possible. Janet Harker established, for example, that cockroaches die of intestinal cancer if internal clocks are

implanted whose rhythm is set in opposite phase to that of
their own internal clocks. This seems to illustrate that an
organism can expire as a result of an unresolved synchronisation problem. According to WARD (1971) Franz Halberg
suggested as early as 1964 that "a connection exists between rhythmic phenomena and cancer". Finally one should
mention the experimental observations of Jürgen Aschoff and
Rütger Wever who found, during their bunker experiments in
Erling-Andechs, that humans who were isolated over a longer
period from every kind of physical time-indication (with
the exception of the time-indication theoretically possible
by gravitational interaction) consistently displayed symp-
toms of a desynchronisation of the sleep-waking rhythm and
the 'temperature-rhythm' if under psychical stress or older
than 40 years of age.[18] Both these features, however, are
indicators for an increased danger of cancer.

It goes without saying that the notion of system-time
can only provide a conceptual framework for the study of
links between rhythms and diseases. However, the close conceptual connection between PEP-systems, system-time, and
system-law necessarily means that certain system-laws which
determine the totality of the system can no longer be valid
if the system-times of subsystems are no longer synchronised.

In general, it can be expected in the field of medicine
that patient-specific lawlikenesses can be discovered if
the system-time of patients are taken into account. For
such investigations to take place system-time clocks would
have to be used whose sole function was to count the periods
chosen as units of time.[19]

If it could be shown, for example, that the pulse-beat,
breathing-rate, or periods of alpha-waves are periodically
equivalent to the activity-passivity rhythm of 9o-1oo minutes discovered by LAVIE (1976) then it would be possible
to construct for human use a system-time clock analogous
to the physical wrist-watch. A full revolution on this
clock would correspond to an entire activity-passivity period, and the individual movements of the hands within this
revolution would be regulated by the pulse, the breathing-
rate, or the alpha-waves, depending on which of these periodic processes were chosen as unit.

Another example of system-time clocks are the revolution-counters of engines or the mileage-recorders of cars. since the mileage-recorder relates directly to the number of times the wheels turn and this number of turns correlates with the revolutions of the engine. At any rate, the revolutions recorded on an engine are a well-known indicator of how long (measured in system-time units) this piece of machinery can be expected to continue to function. This is why the prospective purchaser of a second-hand car (automobile) is interested in knowing the mileage it has done.

With regard to the second empirical question, i.e., that of system-laws, one must take account of the results of investigations discovered in research relating to questions 4.1.1. to 4.1.4. . Here (in 4.1.3. and 4.1.4.) the question of super-laws is raised. We shall call here a super-law the form of a system-law for which it is possible to reinterpret the elements and parameters occurring in that law in such a way that the reinterpretation gives rise to a system-law of another system.

Let us look briefly at the Einsteinian notion of field-equation which is certainly held to be a fundamental system-law of the physical world by the vast majority of physicists today. The construction of this field-equation is carried out in exact accordance with the criteria noted here for a system-law. The system-defining element of energy-mass-distribution is an essential component of the field equation, and the coupling of geometric elements with mass-distribution occurs in such a way that the general principle of relativity (thus also the totality principle introduced here) is fulfilled.

Suppose one had formulated a system-law for a system other than that of the physical world, a system-law constructed in a similar way to the Einsteinian field equation but with the difference that the system-defining elements possess a significance other than the energy-mass-distribution, one could then regard the Einsteinian field-equation as an example of a super-law, and all the structural inferences one can draw from Einsteinian field-equation could be transferred to that other system. The form of the Einsteinian field-equation would thus be a super-law.

In this way, then, super-lawlikenesses would allow the transfer of problems and solutions from one system to an-

other. This is true of course particularly with regard to
problems relating exclusively to the concept of time.

5. CONSIDERATIONS CONCERNING CONSISTENCY

In the second and third sections of this paper I showed
that the introduction of the notion of system-time enabled
the justification of the use of <u>physical time</u> quite naturally in all those cases where it is a question of describing
phenomena of <u>physical time</u>. In this way the arbitrariness
arising from the free choice of PEP-classes for the unity
statement was avoided. At the same time, however, the idea
of an objective, quantitative concept of time had to be
abandoned, if "objective" is understood in the ontological
sense of a "being-in-itself". Thus the distinction between
objective and subjective time also became untenable. Instead
of this, a time measurement can be seen to be adequate or inadequate to its task depending on whether events which can
best be seen as part of a particular PEP-system are described in terms of that system's system-time or with the
time-measure of another system. Thus, for example, it would
not be adequate to mesure micro-physical processes with
Grünbaum's diurnal time T, and even less so with the human
pulse-beat.

It might be objected that physical time is the only
system-time on which subjective agreement can be reached
between individuals. Carnap counters this objection most
strongly when he writes:[20] "Our choice of a process as a
basis for time measurement cannot be seen in terms of right
or wrong. Every choice is logically possible." The theory
of system-times makes it particularly clear that it is indeed practical to use physical time for all statements concerning the physical world.

The question of which system-times are to be preferred
for the description of a process can only be answered by
taking full account of what this description is aiming to
achieve. It might even be convenient to measure the same
process with the aid of different system-times. Thus, when
buying an automobile, both the year of construction (physical
time) <u>and</u> the recorded mileage (the automobile's system-
time) <u>will</u> be taken into account; only when one is in pos-

session of <u>both</u> these items of information can one estimate
the life-expectancy of the automobile.

Proponents of bio-rhythms might object in the following
way to the theory of system-times as it is propounded here:
biological rhythms follow physical rhythms, such as the
earth's rotation on its axis, its orbit around the sun, and
the revolution of the moon around the earth; the synchronisation of the different system-times belonging to one organism probably takes place by means of such physical time-
providers; since higher forms of life are only possible as
a result of this synchronisation, the specific lawlikenesses
of these organisms will themselves follow physical time. It
is hard, certainly, to fully counter this line of argument,
but I should like to suggest the following as worth reflect-
ion. The so-called circadian rhythms were named 'circadian'
by Franz'Halberg precisely because he intended with this
word to express that these rhythms have period-lengths of
<u>approximately</u> (circa) 24 hours. This means that the physical
time-provider offered by the 24 hour light-darkness cycle
leaves the individual organism room for its own cycle, which
does indeed follow the day-night period. And if this is true
for periods of the order magnitude of 24 hours, then it
can certainly be true for shorter periods of some few hours,
minutes, or even seconds. Whether specific lawlikenesses can
be found with regard to this own cycle of the organism can
clearly only be shown by empirical research.

The objections to the theory of system-times become
more critical if one tries to attribute autonomous, biological rhythms to chemical oscillations (c.f., JESSEN (1978)).
And since chemical processes have long been regarded as
special physical processes, biological rhythms would thus
be directly attributed to physical rhythms (bio-chemical
objection).

A corresponding objection to the theory of system-times
could be derived from the general theory of relativity(GTR).
Einsteinian field-equations are in fact fromulated in such
a way as to remain invariant in the face of any and all co-
ordinate-transformations. And this would mean that all the
time mesures I have said to be system-times are in fact already contained in the GTR scheme in the form of special
time-transformations. Thus the concept of time propounded
by the GTR would be the only one conceivable (GTR objection).

Both the bio-chemical and the GTR objections are based on the theory of cognition known as physicalism, i.e., the view that all intersubjectively perceptual phenomena are to be attributed exclusively to physical lawfulnesses.

The physicalist theory of cognition is, however, hard to sustain. This is above all due to the fact that certain fundamental considerations lead one to believe that there will never be such a thing as a complete theory of Physics, and secondly, even were one to consider Physics in its present state as virtually complete, the smooth transition from simple to more complex structures is not feasible because of the vast amount of information involved. It is therefore a hopeless venture for a quantum theorist to wish to calculate quantum-mechanically, say, a DNS molecule, and these difficulties become simply immeasurable if an elementary particle researcher were to attempt the same.

Physicists nowadays have become accustomed to the fact of having to abandon theoretical positions unchallenged for centuries, like for example the postulate of determinism. Some physicists, inspired by Einstein's theories, tried again and again to attribute quantum theory – the enfant terrible of Physics – to a deterministic theory, though as yet without success.[21] If one wishes to continue believing in a deterministic theory, more complex structures will have to be found than has up till now been the case.

The theory of system-times has the advantage over the physicalistic evaluation that its simple class-formation allows one to arrive at clearly visible PEP-systems intended to display simple lawlikenesses in relation to the relevant system-time. To try to attribute such lawlikenesses – should any in fact be found – to a physicalistic point of origin would be comparable to attempting to deduce quantum theory from a deterministic theory.

If then the theory of system-times puts paid to the possibility of introducing an objective, ontologically distinct time-measure, how can one best understand within the framework of system-time theory the widely held conviction that everything that exists is temporally determined, or further, that time is a constituting attribute of reality? To put this question another way, if reality is something objective in itself but metric time does not in itself represent an objective element, to what extent, then, can time be part of reality?

This question is not as problematic as it might appear at first sight, for the theory of system-times assumes a <u>given</u> topological notion of time, and most of the ideas of time we gather from everyday experience are tied to this topological notion of time, above all, for example, the idea that whatever occurs can be ordered into a sequence of earlier and later. The scheme of system-times thus still permits the ontological conception of time as a ceaseless, flowing stream which carries all that is real along with it, for it is quite possible to be of the opinion that the reality of time does indeed exist, but that theoretical reasons alone prevent one from discovering an objective measuring procedure for this time. If one recalls, though, that the notion of system-time is in fact system-constituting and thus itself first creates the framework and even the objects of possible (temporal) experience, then this line of argument runs very close to the Kantian idea that time has no significance in itself but is to be understood as belonging to the necessary conditions of conceivable experience; there is clearly, though, the difference that for KANT (1787) only <u>one</u> time-scale was conceivable, as indeed there was only <u>one</u> space, the Euclidean, whereas here an infinite number of time-scales can be defined, even such that do not allow of mutual transformation.

Despite its proximity to Kantian formulations, the notion of system-time does in fact have a strong empirical side to it. This is seen in the fact that the PEP-classes have to be discovered empirically, and that it is an empirical statement when we are able to say that in relation to a given degree of accuracy there are indeed many non-trivial PEP-classes. It remains unclear whether all PEP-classes might not perhaps split up into subclasses finally containing only one element if the degree of accuracy is continuously raised. In this case we could no longer speak of the reality of metric times, and from this it becomes clear that the search for the reality of metric times is at the same time the search for those connections between processes which ensure the existence of non-trivial PEP-classes. As long as the principle of action by contact is maintained, one can imagine that such connections are effected by means of media. Thus, in the example already mentioned, in cock-

roaches the connection-forming medium is provided by the
cockroach's blood.

Since the assumption of an ether has become otiose
since Einstein's relativity theory, it is not at all easy
to say how one should best think of the connection-forming
medium in the PEP-system of the physical world. Ernst Mach
defined physical time as "the dependence on one another
of changes"[22] , and he marvelled at such dependencies when
he wrote: "During the time in which, for example, the earth
completes 1/86400 of its rotation and covers a corresponding fraction of its orbit, light travels at the same time
a distance of 3oo.ooo kilometers, a body in free fall falls
4.9 meters downwards, a thread pendulum almost 1 meter long
completes a single oscillation, and every thermal, electro-
magnetic, chemical process proceeds to an extent determined by the circumstances of its environment but also by
the fraction of the earth's rotation. Does this not strike
one as very remarkable? What have all these processes to do
with the earth's rotation?"[23]   One could as well ask, what
connection-forming medium forces on them the same step?
Mach does not try to conceal the fact that the only answers
he can find are unsatisfactory, and he conjectures that a
"more penetrating insight into the interrelation of bodies"[24]
could discover the mutual dependence. It is Mach's view
that this connection must be searched for, since, as he
writes, "if one allows oneself, without striving for such
an insight, to trust as fact one's impression that even the
satellites of Jupiter keep pace with physical processes on
the earth, then one is not all that far from the mystical
ideas of the astrology of the Middle Ages."[25]

In my view Mach was already thinking here along the
lines of a theory of system-times, in that he was searching
for a reason why the assumed proposition of congruence can
be used for the physical world. In a similar way he had previously searched for a reason for the inertia of masses,
and by taking account of such considerations Einstein was
then able to produce his general theory of relativity. This
leads one to suspect that one should search for the necessary connection-forming medium for physical time within the
framework of the general theory of relativity. This Einsteinian picture of the physical world enables connection
in the physical world to be established by means of the so-
called affine connection of Riemannian geometry. Affine

connection, which arises out of Riemannian metrics,[26] ties
together the tangential spaces of the underlying (space-
time) manifold, so that vector fields can be defined in the
tangential spaces, thus making Physics, in the form of
field-Physics, possible. Connection in the physical world
is thus here guaranteed by the postulate that it is de-
scribable in terms of Riemannian geometry. The only possible
answer within the framework of the GTR to Mach's question of
why there is in the physical world a temporal pace-keeping
of spatially separate progressions: because we can use
a Riemannian space to describe the physical world. And it is
of no importance at all what form the Riemannian metrics
take, the only important point being that a Riemannian metric is assumed. In this way we have almost returned to
the Kantian position of believing that space and time are
the pure forms of visualization and are thus conditions for
any and all possible physical experience. What is empirical
within the framework of the GTR is first of all that precisely a 4-dimensional Riemannian metric must be assumed,
and secondly that this metric has a particular structure.

We cannot, therefore, speak of a connection-forming
medium within the framework of the GTR. However, this is
to be expected from a theory which - like the GTR - fulfills the totality principle. For in order to decide whether
a medium has or has not the property of forming connections
one needs a position external to these connections, and this
is conceptually impossible within the framework of the GTR.
If with regard to cockroaches it was recognised that blood
was the medium which made the cockroach into a PEP-system,
this was only possible by regarding the blood of the cockroach as a component of the physical world.

If, then, a reason is to be found for connection in
the physical world, this reason will only be stateable if
one succeeds in integrating the physical world into a more
comprehensive system.[27] This is a question which must also
be confronted when thinking of the cosmological problem of
the beginning and the end of the physical world. Within the
framework of time metrization procedures physical time is
only definable if the physical world exists. If the physical
world arose out of the so-called 'Big Bang', then the question of "what came before" has no physical meaning. The same
is true, at the other end of the scale, if the physical world
is disappearing into a black hole. This holds true in the same

way for all PEP-systems: if one is concerned with the genesis or expiry of a system, then there is no 'before' and
no 'after' in relation to the relevant system-time.

Every system exists eternally in relation to its own
system-time.[28] This means that within his own system-time
each human being is immortal. If only my own time has any
importance to me then it is as senseless as in the cosmological case to ask about 'before' or 'after'. Epicurus
seems to have made this inference when he writes in his
letter to Menoikeus:[29] "Accustom yourself to the thought
that death does not concern us at all" and justifies this
when he continues: "for as long as we exist, there is no
death, and when death is there, we no longer exist."[30] Nevertheless the notion of death and the fact of speaking about
future non-existence are only possible by virtue of the
idea that we live in a wider system, for with regard to
physical time I have a definite birthday and will have a
definite day on which I die. By extension, one can ask with
regard to the physical world whether another system than physical time is conceivable in which it makes sense to talk
of the beginning and the end of the physical world. Anyone
who does not regard as meaningless the talk in the previous
sentence of the beginning and the end of the physical world
should therefore regard the existence of that other wider
system as possible.

NOTES

+ This paper is a translation of a slightly altered version
 of Grundlagen einer Theorie der Systemzeiten, Allgem.
 Zeitschr. f. Philos. Vol.6/2 (1981)

1. Generally speaking, both mapping d and domain D will
 have to meet certain further conditions if we are to
 talk of metrization. The problems concerning this are
 not entered into in this paper.

2. It has been suggested also that no periodic process is required for the unity statement, e.g., JANICH (1969) or
 GONSETH (1971). But the arguments against a TMP with means
 of periodic processes are not convincing.

3. CARNAP (1966), German edition, p. 87. I thank P. Janich
 for his information that this idea stems from M. Schlick.

4. It should be noted that without a minimum principle of
 induction, the notion of periodic equivalence cannot
 be introduced.

5. c.f., MERCIER (1975), p. 532.

6. c.f., for example, CARNAP (1966) and STEGMÜLLER (197o).

7. In more recent literature on cosmology this assumption
is also referred to as unity principle. See, for example,
WEIDEMANN (1978). In this paper we read of the unity
principle: "it ties everything to everything". (p. 686)

8. This definition of the physical world has the remarkable consequence of making it doubtful whether individual quantum-physical events such as the decay of the
nucleus of an atom can be counted as belonging to the
physical world. According to quantum mechanics, the
conditions pertaining to the decay of a single nucleus
cannot be fully described, or are purely fortuitous, and
in this last case the above-mentioned consequence would
arise. This leads one to think that quantum-mechanics
deals with different times than the physical ones defined here.

9. This concept of system-time should not be confused with
the synonemous notion proposed by F. Adler. The latter
is only a special case of the concept defined here. c.f.,
ADLER (192o).

1o. I am aware that the isolation of the concept of time
from the concept of space is problematic. However, in
more recent work concerning the space-time problem, attention is drawn precisely to the special position of
the concept of time, and the total fusion of this concept with the concept of space is rejected. c.f., MISNER
(1973), HÜBNER (1978) or PROKHOVNIK (1976).

11. c.f., GRÜNBAUM, pp. 66-77.

12. c.f., EINSTEIN (1969) p. 59, or c.f. in a very early
formulation EINSTEIN (1918).

13. In order to understand better what 'coordinates' or a
'descriptive base' within a PEP-system mean, it is clear
that the concept of system-space must first be defined
more closely.

14. As is standard usage in the philosophy of science and
logic, "essential" is taken here to mean that an element
occurs essential in a proposition when it is possible to
exchange this element with another in such a way that
the truth-value of the proposition is changed.C.f.,STEGMÜL-
LER (1969), S. 3.

15. See here STEGMÜLLER (1969), ch. V.

16. c.f. WARD (1971) p. 174 (German ed.).

17. Among periodically equivalent processes one can distinguish between processes which are dependent on other

processes with regard to their periods, and processes
which do not display this dependence. To the latter belong those processes which we call biological clocks,
such as, for example, the neuro-secretory cells discovered by Janet Harker in the lower esophegeal ganglion
of the cockroach.

18. c.f., WEVER (1976).
19. Theoretically all investigations to discover system-
laws can also be carried out with the aid of physical
time. If, parallel to measuring the system-element in
question, one measures with the same physical clock the
periodic process to be used for the time metrization,
then it is not difficult to discover the dependence of
the system-element on the system-time by eliminating the
physical time variable. (I am indebted for this insight
to R. Werth, with whom I discussed this point at the
World Congress of Philosophy in 1978.
2o. c.f., CARNAP (1966), German edition, p. 91 (my translation).
21. c.f.,HÜBNER (1978), ch. II and VI.
22. c.f., MACH (191o) p. 495 (my translation).
23. ibid, p. 495 (my translation)
24. ibid, p. 496 (my translation)
25. ibid, p. 496 (my translation)
26. The affine connection of Riemannian geometry is distinguished by the so-called Christoffel symbols Γ^i_{jk}. These
arise by means of partial differentiation from the metric tensor (g_{ik}):

$$\Gamma^i_{jk} = \frac{1}{2} g^{iv}(g_{jv,k} - g_{jk,v} + g_{vk,j}) ,$$

where the summation over the same indices is assumed
from 1 to 4 and the last character within the brackets
design the coordinate of partial differentiation.

27. I have already indicated in footnote 8 that in the definition given here of physical world, quantum-mechanics
should be understood as referring to another system.Since
in quantum field-theory the vacuum - thought of as in-
finitely many oscillators in their ground state - could
be seen as a medium providing through the identical
rhythm of all oscillators the reason for the temporal
pace-keeping of spatially separate progressions, then
in accordance with our considerations, quantum field-
theory would also refer to a larger system than the GTR.
The first line of thought which occurs is to extend the
physical world by means of all unrealised possibilities,

which is e.g. approximately to Wheeler's framework of superspace (c.f., MISNER (1973) or WHEELER (1979).

28. The notion "eternal", understood as "without beginning and without end", is reinterpreted with this sense here to apply to system-times.

29. c.f., EPICURUS (1949) p. 45 (my translation).

3o. Prof. José Echeverria drew my attention to this connection with Epicurus during the symposium held by the Greek Humanist Society in Greece in September 1978.

REFERENCES

Adler, F.: 192o, Ortszeit, System-Zeit, Zonenzeit, und das ausgezeichnete Bezugssystem der Elektrodynamik, Wien.

Carnap, R.: 1966, Philosophical Foundations of Physics,New York. German edition: Einführung in die Philosophie der Naturwissenschaft, München 1969.

Einstein, A.: 1918, Prinzipielles zur allgemeinen Relativitätstheorie, Ann.d.Physik 55, pp. 241-244.

Einstein, A.: 1969, Grundzüge der Relatiyitätstheorie, 5th ed., Berlin.

Epicurus: 1949, Von der Überwindung der Furcht, German translation and with a preface by O.Gigon, Zürich.

Gonseth, F.: 1971, From the measurement of Time to the Method of Research, in: Zeman, J. (ed.): Time in Science and Philosophy, Amsterdam, pp. 277-3o5.

Grünbaum, A.: 1973, Philosophical Problems of Space and Time, 2nd ed., Dordrecht-Boston, Boston Studies in the Philosophy of Science, Vol. XII.

Hübner, K.: 1978, Kritik der wissenschaftlichen Vernunft, Freiburg.

Janich, P.: 1969, Die Protophysik der Zeit, BI-Hochschul-taschenbücher No. 517, Mannheim.

Jessen, W.: 1978, Chemische Oszillationen und Strukturen als Grundlage einer zeitlichen und räumlichen Organisation, Naturwissenschaften 65, pp. 449-455.

Kant, I.: 1787, Kritik der reinen Vernunft, Riga.

Lavie, P., and Kripke, D.F.: 1976, Bio-Rhythmen 11, die 9o Minuten Uhr in uns, Psychologie heute, Juni 1976, pp. 6o-63.

Mach, E.: 191o, Popularwissenschaftliche Vorlesungen, 4th ed., Leipzig.

Mercier, A.: 1975, Epistemological Questions concerning Cosmology and Gravitation, General Relativity and Gravitaion 6, pp. 513-536.

Misner, Ch.W., Thorne, K.Th., Wheeler, J.A.: 1973, <u>Gravitation</u>, San Francisco.

Prokhovnik, S.J.: 1976, Time as a Universal Property of Nature, Preprint.

Reichenbach, H.: 1928, <u>Philosophie der Raum-Zeit-Lehre</u>, Berlin and Leipzig.

Stegmüller, W.: 1969, <u>Probleme und Resultate der Wissenschaftstheorie u. anal. Philos.</u>,Vol. 1: <u>Wissenschaftliche</u> Erklärung und Begründung, Berlin.

Stegmüller, W.: 197o, <u>Probleme und Resultate der Wissenschaftstheorie</u>, Vol. 2: <u>Theorie und Erfahrung</u>, Berlin-Heidelberg-New York.

Ward, Ritchie R.: 1973, <u>The Living Clocks</u>, New York 1971, . German translation: <u>Die biologischen Uhren</u>, Reinbek.

Weidemann, V.: 1978, Cosmology: Science or Speculation? in: <u>Sektionsvorträge des 16. Weltkongresses für Philosophie 1978</u>, Düsseldorf, pp. 683-686.

Wever, R.: 1976, Probleme der zirkadianen Periodik und ihrer Störungen, <u>Arzneimittelforschung</u> 26, pp. 1o5o-1o54.

Wheeler, J.A.: 1979, Frontiers on Time in: Toraldo di Francia, G. (ed.) <u>Problems in the Foundations of Physics</u>, Proc.Int.Sch.Phys., "ENRICO FERMI" LXXII 1977, Amsterdam.

ACKNOWLEDGEMENTS

I am especially grateful to Professor Dr. K.Hübner for his critical advice and encouragement. For their inestimable help in translating and typing the manuscript I should like to mention Timothy Martin and Brigitte Scholz.

Peter Janich

NEWTON AB OMNI NAEVO VINDICATUS (1)

Two theories, Euclid's geometry and Newton's mechanics,
have persistently influenced the history of the exact
sciences for centuries. Common to both is that their historical and systematic impact has been due not only to their
outstanding positive achievements, but also to their specific
deficiencies. As is well known, Euclid's geometry, which
is oriented upon the Aristotelian ideal of theory, begins
with a series of definitions which are a) insufficient from
the modern logical standpoint or the standpoint of the
theory of definition; (and which b) play a role neither in
the proof of the theories, nor in the propositional content
of the theory. Gauss was not the first to find (2) with
the absence of (in modern terms) axioms of order - thus,
essentially, of rules for the use of the word 'between' - in
Euclid. It can be shown historically that the logical weak-
ness of these definitions, as well as their systematic insig-
nificance for geometry resulting from these weaknesses, have
led to a twothousand year history of attempted repairs and,
finally, to the modern, formalistic conception of geometry
with D. Hilbert. According to this conception, the defining
of fundamental terms was banned from mathematics by programm-
atic decree. (3) Many mathematicians, as well as philosophers
of science have, somewhat too quickly, coupled this conception
together with the notion that the basic concepts of geometry
are not definable because they have not yet been defined.

Newton's mechanics in the <u>Principia,</u> which in its
organizational form is orientated upon the example provided
by Euclid's geometry, is aquainted with a similar problem:
The definitions at the head of the laws of motion and their
elucidations had (due to deficiencies to be discussed below)
historical consequences for the development of a modern
concept of theory in physics analogous to those just mentioned regarding Euclid's geometry. At issue here are not so
much the well known definitional problems pertaining to
absolute space and absolute time as those relating to the
dynamical terms; namely, mass and force.

The present article makes clear, first of all, that
there is a conspicuous discrepancy between historical and

225

D. Mayr and G. Süssmann (eds.), Space, Time, and Mechanics, 225–240.
Copyright © 1983 by D. Reidel Publishing Company.

sytematic evaluations of Newton's inertial mechanics. This
discrepancy derives from the following state of affairs:
Whereas historians of science see in the <u>Principia</u> the
culmination and preliminary termination of a process of
mechanics' development leading to Newton's law of inertia,
from a systematic perspective this particular principle of
inertia is less satisfactory than earlier versions. Secondly, the modern-day evaluation of the <u>Principia</u> and, furthermore, of the basic ideas of inertial mechanics are treated
of to the extent that the consequences for the modern concept
of theory in physics which are normally drawn from this
evaluation, can be made clear.

1) On the development of the classical principle of inertia.
One may summarize how the prehistory of the principle of
inertia is nowadays written in the following way:
In his theory of motion in the <u>Physics</u>, Aristotle asserted
that everything that moves must be moved by something and
that a movement will cease if its cause is no longer present.
In the special case of vertical motion, the cause of motion
for Aristotle is the respective natural place of the four
elements - i.e., earth, water, air and fire - in their striving
to arrange themselves in the form of concentric spheres around
the centre - point of the earth. Correspondingly, the material
composition of a vertically moved body determines whether
this body will move upwards or downwards. In the modern way
of speaking, aristotelian principles are appropriate for a
consistent explanation of vertical movements in terms of
gravity and buoyancy.
      Now, Aristotle differentiates between 'natural' and
'artificial' motion according to whether moved bodies contain
whithin themselves the cause of their motions, or not. In
keeping with this, he subsumes vertical motion under natural
motion. On the other hand, motions of thrown bodies, or
eventually, the motion of a wagon pulled by a horse are
cases of artificial motion. In order to account for throw
in a way consistent with the principles cited above,
Aristotle is forced to resort to a notion that has always
been found to be rather implausible. It is that the air
surrounding a thrown body pushes it onwards after its leaving
the hand of the thrower. This unconvincing supposition
(unconvincing, insomuch as it is not based upon an appeal
to everyday experience) provoked the criticism of Aristotle's
successors. Beginning with Philoponus (4) it provided the
incentive for the formulation of competing 'impetus theories'

for the explanation of trajectories. The manifold types of
'driven', or in modern terms, of 'forced' motion were largely
forgotten in the course of this dispute.

The core of impetus theories, which were brought to
their full bloom during the fourteenth-century by John
Buridan of the Parisian Terminist school, is the notion that
bodies in motion have an 'impetus', dependent upon velocity
and size and which is annihilated by air resistence and
gravity. By way of Albert and Saxony, Nicolas of Cusa, of
Buonamici - the teacher of Galileo -, of Benedetti and,
finally, of the young Galileo himself, impetus theories were
extended to include attempts at explaining free fall. They
were, thus, carried into the immediate proximity of classical
physics, modern philosophers of science, however, locate the
point of differentiation between scholastic and inertial
mechanics in the formulation of the law of inertia at hand
in Galileo's Discorsi (1638). In the part of this work
entitled "The Third Day" there is a scholium in which it is
discussed whether a body rolling down an inclined plane
obtains velocity to allow it to ascend on another inclined
plane to the height of its point of departure. In the modern
way of speaking this question concerns the problem of the
conservation of energy. (5) In this context, motion on a
horizontal plane is claimed to be a limiting case between
the positively accelerated motion of descent and the nega-
tively accelerated motion of ascent. A body moving horizontally would maintain without end the velocity attained at the
lowest point of the descending movement. (6) The condition
of frictionlessness, which is presupposed here, is expressly
mentioned by Galileo in preceding passages.

Modern historians of science correctly see contained
in these tenets of Galileo the basic conception of the
principle of inertia. To adopt a suggestive formula: If
forces were regarded by Aristotle as cause of velocity, for
Galileo they serve only in the explanation of change of
velocity. Frictionlessness presupposed, constant velocities
require no forces for their explanation. Modern literature
on the subject normally assesses as a kind of blemish upon
the Galileon law of inertia of the fact that the continuation
of horizontal motion on the earth results in a circular orbit.
Hence, the formulation of the law of inertia which is, within
the scope of classical physics, valid to this day is to be
discerned in the first law of motion in Newton's Principia,
where (a) horizontal motion is replaced by rectilinear motion;
(b) the earth is stipulated as the frame of references for

the motion in question; and where (c) the state of rest is
seen to be on an equal footing with motion of constant
velocity. Already in 1656 - i.e., thirty years prior to the
publication of the Principia - Christian Huygens began his
treatise, De motu corporum ex percussione, with the
following hypothesis: "Corpus quodlibet semel motum, si
nihil obstet pergere moveri eadem perpetuo celeritate et
secundum lineam rectam". (7) Apart from the explicit mention
of the state of rest, this is the same statement as can be
found in Newton.

Nowadays we see one of Newton's signal achievements
to consist in the fact that he succeeded in "applying
terrestrial mechanics to the heavens " - i.e., that he
recognized that the laws of fall and throw are equivalent
to Kepler's laws of planetary motion. In this way Newton
was able to comprehend the motion of the planets as 'free
fall' within the gravitational field of the sun and to explain
their laws of motion in terms of a law of gravity. For modern
historians of science, this appraisal of Newton's achievements implicitly contains the danger of interpreting earlier
theories of motion as more prehistory of inertial mechanics.
This, of course, is not to say that the fraternity of historians of science has not attempted to understand pre-Newtonian
theories within the context of their own epochs and historical presuppositions. But they do let it be known that Newton,
in his formulation of the pertinent principles, successfully
concluded a centuries - long controversey over the correct
theory of the motions of fall and throw (8). In particular,
the prehistory of mechanics and (to quote a chapter heading
from E. Mach) the development of the principles of statics
(Die Entwicklung der Prinzipien der Statik) remain separate
from one another.

This assessment on the part of the history of science - which gives preferences to Newton's principle of inertia,
due to the greater generality of the former - is opposed by
a systematically oriented evaluation. Subject to discussion
here is, first of all, the above cited claim of Galileo at
hand in the Discorsi. Reformulated in modern terminology it
would run approximately as follows: "if a body is moving
without friction on a horizontal plane, it will maintain
constant velocity.

In order to answer the question of these status of this
principle - i.e., whether we are dealing with a definitoric,
an empirical, or perhaps a senseless proposition - we need
to ask whether, and to what extent, the components of this

proposition are definitorically fixed by Galileo. The word
'horizontal' does not provide any difficulties here, as it
can be fixed by example with reference to water surfaces
at rest. Constant velocity is explicitly defined in the
Discorsi by referral to the condition that, in equal times,
equal distances must be covered. (9) Galileo does not speak
of the measurement of lenght; for this seemed to be something
that one could take for granted, given the practices of
measurement in handicraft. As is well known, Galileo
accomplished the measurement of time first by counting his
pulse and later by use of a primitive water-clock (10).
Regarding the question of completeness of the definition,
it is significant that, in measuring time, Galileo does
not have recourse to inertial motion. Hence, it can be maintained that constant velocity is also explicitly and operatively determined in the Discorsi. In light of this, the
only undefined expression in Galileo's law of inertia remains
the word 'frictionless'. From the modern point of view,
therefore, this law of inertia can be regarded as an operative definition of frictionlessness that is in no sense
arbitrary. Rather, it agrees eminently with everyday experience,
according to which a ball will roll a greater distance on a
horizontal plane relative to a higher degree of smoothness
of that ball and of the surface of the plane. One may not,
of course, suppose that Galileo himself meant his law of
inertial to be a definition of frictionlessness. We are only
concerned with the supposition that this law, when comprehended
from the modern point of view, can be read as a definition
against which no further logical objections, or objections
stemming from the theory of definition can be raised.

The matter is different with respect to Newton. As
expressed in the first law of motion (11), the principle of
inertia is neither an empirical proposition, nor – seeing
that no appropriate frame of reference is indicated – a
definition of the absence of external motive forces. The
customary discussion, in which it is demonstrated that this
deficiency cannot be overcome even with the aid of the
other two laws of motion, will not be repeated or recapitu-
lated here. (12) Even if one assumes that Newton had in mind
a method of time-measurement available independently
of the laws of motion of his mechanics, the most favourable
interpretation (i.e., that, in the first law of motion,
appropriate frames of references are defined by the absence
of forces; and that, in the second law (13) forces are
defined with regard to the appropriate frame of reference)

leads merely to a definitoric circle. Without supplementation,
the mechanics expressed in Newton's laws of motion runs into
a substantive contradiction, namely, that a frame of reference
is to be defined dynamically (i.e., via a dynamical link
between bodies) through isolation conditions (i.e., without
a dynamical link. But, of course, modern-day analytic philosophy of science argues logico-syntactically, instead of
with regard to the contents.

2.) Modern evaluations of Newton's principle of
inertia.
A comprehensive resumé of the modern-day discussion of
definition problems inherent in Newton's mechanics is to
be found in the work of W. Stegmüller. (14) In a volume
written before Stegmüller's 'Kuhn-Sneed-Turn' the difficulties just dealt with are considered under the title,
"Variable Possibilities in the Interpretation of Theories"
("variable Deutungsmöglichkeiten von Theorien"). The outcome of this chapter is approximately as follows; For all
propositions of mechanics, only those propositions can
be said to be empirico-hypothetical, for which all expressions contained in them are operationally defined beforehand.
However, (and this hold for the law of inertia as well) if
all expressions of a proposition are determined except for
one, then that proposition is to be understood as a convention, or respectively, as a definition for just that undefined expression; and for this, all expressions contained
in the law of inertia, taken singly, are at issue. Further
analysis makes evident that for the avoidance of ambiguity
it is not permissible to ignore the problem of appropriate
frames of reference. In light of this, Stegmüller can provide
an evaluation of Newton's definitoric problems that agree
with the authors he refers to (15). This evaluation amounts
to the following two assumptions:
a) Whether the law of inertia is an empirico-hypothetical
assertion of a fact or a convention depends, if at all
determinable, upon the status of other theorems of mechanics.
Unequivocal classification is impossible; singular propositions of the theory remain variably interpretable.
b) Taken in isolation, the law of inertia is senseless.
Only a "theory as a whole" has factual content (p. 124).
      This conception is significant primarily because, by
means of it, explicit conclusions concerning modern-day
understanding of theory can be drawn from Newton's definitional dilemma, or from its interpretation. According to

that understanding, axiomatically formulated theories of
physics make sense only when taken as wholes and are to be
regarded formally. Furthermore, at least partial definitions are to be sought by way of correspondence rules,
interpretations, operational definitions and the like. This
conception, which stems from analytic-empirical philosophy's
theory of the language of science and which is obliged mainly
to Carnap, Hempel and Nagel, has found further development
in Ramsey-solution of the problem of theoretical terms.
Concerning this method, it is to be noted that Stegmüller
maintains that for every axiomatization of classical mechanics the non-definability of force and mass (and, thus,
the insolubility of Newton's definitional dilemma) can be
stringently proven. (17) Of course, such a proof, which
adheres to the theorem of Padoa, is valid for only those
theories, whose vocabulary is supposed to be (implicitly)
determined by axioms alone. (18)

For the problematic analysed here, it is significant
that Newtonis nowadays read by analytic philosophers of
science in such a way that reasons formalistic understanding
of theory are inferred from the non-controversial definitory
flaws in Newton's inertial mechanics. (19) It is generally
noted merely in passing, if at all, that the definitions
preceding Newton's laws of motion (such as that of mass as the
product of density and volume) are, in an entirely common-
place sense circular or otherwise insufficient. Analogous
to the definitions preceding Euclid's axioms, Newton's
definitions are normally seen as irrelevant - indeed, as
well meant protreptically, but, in the final analysis, as
mere introduction burdened with philosophico-phenomenological
ballast. According to the contemporary 'non-statement view'
so highly esteemed by analytic philosophers of science, it
remains only to be asked whether Newton's mechanics, now
seen as formal theory-structure, can by some stroke of luck
find a model (in the sense of having an application). One
does not run into contradiction with these conceptions in
maintaining, that there is no single, conclusive argument
for the move, made from the non-definability of the terms
'force' and 'mass' to non-definability. (20)

3. Newton's laws of motion without definitional
gaps.
The purpose of the following is to substantiate the claim
that Newton's mechanics (as articulated in the Principia)
can be interpreted in such a way that, with the aid of

several additions compatible with the text, it fulfills
modern-day requirements of definitional completeness and
clarity with respect to the status of individual propositions.

Without any further introduction (apart from a preface)
Newton begins the Principia with the formulation of a
definition. According to this definition, the quantity of
matter is measured by taking together density and volume.
(21) In the subsequent elucidation it is stated that the
quantity of matter is, in addition, to be understood as
'body' or 'mass' and is known by the weight of a given
body. (22)

For the modern reader, who, as a result of his schooling
in physics, is used to defining density as the quotient of
mass and volume, Newton's definition appears rather odd. For
us - all of whom are anchored within the Newtonian tradition
- mass is, in spite of the definitional deficiencies discussed
above, the familiar and, therefore, elementary concept. But,
obviously, Newton's own view of the matter is just the
opposite, and indeed, everyone is aquainted with a pre- or
extra - scientific, elementary (if not necessarily metrical)
concept of density, to speak somewhat imprecisely, everyone
knows that "lead is heavier than water". Moreover, the
determination of density by means of the buoyance of standard
floater (e.g., in determining whether radiator fluid in a car
is winter-proof or in finding out specific weight of new wine)
is nowadays part or ordinary experience. The production of
bodies of homogeneous density in metal refining processes,
for instance, had by Newton's time long been practically a
matter of course. And in any event, liquids provide, so to
speak by nature good examples of homogeneously dense bodies.
Their densities can often easily be registered by means of
greater or lesser buoyancy; for instance, when, in an
oil-water mixture, the oil accumulates after a certain
period of time in a layer on top of the water. (23)

What Newton does not explicitly discuss is the question
of how density becomes, in modern terminology, a metrical
concept; that is, how at least ratio between densities is
assigned a number means of some procedure of measurement.
In the knowledge of the definitional problems concerning
force and mass, the task is now to explicitly determine a
concept of density in such a way that the solution of the
said definitional problems is not presupposed. Specifically,
the definition of density must be independent of the
determining of inertial frames of reference. Apart from
nineteenth century insight into the logical structure of

measured quantities, this task is soluble by means which
were, without exception, already available to Newton.

One of the oldest instruments of measurements development by mankind is the symmetrical beam balance. Symmetry
conditions can be defined by use of geometry alone. For
purposes of definition, the norm can already be formulated
here that equality in weight ought to be a equivalence-
relation. This norm is fulfilled in the factual use of
balance-scales both outside of and within the sciences.

Now, in the case of the symmetrical beam balance,
everyone understands 'balance' in such a way that, in the
state of equilibrium the beam lays horizontally and the
balance is at rest relative to the earth. A nonvertical
application of the same symmetrical arrangement for the
allocation of tractive power to two beams of equal length
takes place in traction devices, where two draught-animals
are harnessed to a wagon or plow. This arrangement had
also been in use for centuries by Newton's time. One borrows
from this apparatus when a generalized utilization of
balance-scales is recommended for the preparation of the
definition of mass in Newton's sense. In order to achieve
independence of the reference-frame, earth, balance-scales
should be able to be utilized in arbitrary motion relative
to the earth. In the generalized utilization of balance-
scales (i.e., where these also may be accelerated relative
to the earth), two bodies may be called 'equal with respect
to traction', in short, t-equal, when they evoke no rotation
of the balance beam relative to the direction of traction of
the suspension. Then it can be required of such balance-
scales that equality of traction should be an equivalence-
relation. (24)

Such a norm of instrumentfunctioning affords an operative definition of a two-value predicate for bodies; the
formulation of such a norm implies that, in the case of
disturbation (e.g., of symmetry), causes are to be looked for
empirically. In this way it can be shown, that 'disturbing
circumstances' such a buoyance of friction of the surrounding medium can be recognized and rectified by improvement
of the arrangement of the balance (for instance, by under-
taking the comparison in an evacuated container or by
producing equality of friction - these being, in turn,
determined solely with the aid of symmetry and transitivity
of equality of traction, in short, t-equality. (25).

'Homogeneous density' can be defined with the aid of
t-equality: a body is homogeneously dense, if any two of

its parts, which are of equal volume, are t-equal. Thus,
it is with the aid of t-equality that homogeneous density
is defined by means of the measurement of volume; and,
hereby, it is at no point necessary to distinguish a definite
- and indeed: inertial frame of reference. The density -
ratio of homogeneously dense bodies can be determined by
means of the ratio of the volumes of two t-equal parts of
these bodies.

 The mass-ratio is, of course, of greater importance
for the reconstruction of Newton's definition of mass. The
following definition is, therefore, recommended: Two
(arbitrarily selected and, hence, not necessarily homogeneously
dense) bodies have a mass-ratio which is numerically equal
to the ratio of the volumes of two parts of a homogeneously
dense body, these being t-equal to those two bodies. For
illustration, bodies, whose mass-ratio is to be determined,
are each balanced with a homogeneously dense body, for
instance, water, and then the volumes of water are compared
with one another. Because, in making measurement possible, one
is only concerned to indicate a procedure that results in the
determination of ratios - and, hence, because the definition
of a unit of mass in the sense of universal reproducibility
is an entirely different task - it can be maintained that
a explicit definition of mass in Newton's way is fixed when
an operative definition of density like given above precedes
it. In such a way a concept of mass is made available independently of an axiomatically understood mechanics (taken in
the modern sense) and which allows for a non-circular reading
of Newton's three laws of motion:

 To begin with, an appropriate frame of reference for
the laws is to be determined. A methodical choice needs to
be made here, for which different possibilities are open
in dependence of different purposes. If, due to technical
interests and building upon every-day experience with heavy
bodies of little friction, one endeavours programmatically
to measure only 'accelerating' forces - if, in other words,
Newton's second law is to be made into a definition of force -
then it is recommended that one distinguishes between frames
of reference by asserting that no accelerations can occur
without the action of external forces upon a body. This
way of speaking is, of course burdened with the same
definitional problems as appear in the traditional way of
reading Newton. Nevertheless, these problems can be avoided,
if one assumes, for instance, that t-equal bodies have
the same effect upon each other in (central, non-elastic)

collisions. (26) On the supposition that this way of talking
about 'effect' can be made up for terminologically by the
definition yet to be formulated, then Newton's third law
(27) can be read as a mere terminological rule of symmetry
for the direction and magnitude of two bodies' effect upon
each other. This rule of symmetry would, in turn, be justified by the methodological resolution to fix comparisons of
bodies with respect to geometrical, cinematical, or dynamical
criteria in such a way that neither of the two can be marked
as absolute reference-body.

Correspondingly, frames of reference are distinguishable on the condition that, within them, pairs of t-equal
bodies in central, non-elastic collision are indistinguishable.
That is to say, explicitly, that these bodies come to rest
upon any coordinate axis, if they have equal and opposed
velocities before collision.

In accordance with this distinction of frames of reference,
it makes sense to read Newton's second law of motion as a
definition of force, whereby the quantity of motion, taken
as the product of mass and velocity, is already operatively
determined and available. Finally, under these circumstances
the first law of motion - thus, the principle of inertia -
is a logical implication of the second law; for, on the
presupposition of the constancy of mass, the first law follows
directly from the proportionality of force and change of
velocity.

Thereby, it can be maintained that, by giving a definition of density first, Newton's mechanics is satisfactorily
interpretable for the modern-day reader. And with this, the
dispute about the definition of force, mass and inertial
system that has lasted for over two-hundred years may be
regarded as settled.

In conclusion, we shall again consider those consequences
for the concept of theory in modern physics that are drawn
from the previous way of reading the Principia:
As has been shown, the supposed definitional deficiencies
in Newton's mechanics is owing to the resolution to comprehend axiomatic systems in physics solely as propositional
forms, or as formal structures that are to be interpreted
afterwards. Implicit in this resolution, which is connected
with the formalistic turn in mathematics, is the decision
not to take seriously definitions that precede axiomatic
theories. The claim that theories of physics, because of
well known definitional problems of mechanics, ought to be
axiomatic represents a restriction upon scientific discourse,

a restriction which draws its legitimation just by interpretation of older texts under this restriction. In other
words, this restriction is not founded by arguments. In any
event, Newton's mechanics can no longer be seen to provide
reasons for justifying the absence of explicit definitions
in physical theories - no matter how roundabout this justification might be, by way of the most excating theories of
scientific language and by accounts of physical theory
based upon a theory of models.

FOOTNOTES

(1) Gerolamo Saccheri's treaties, Euclides ab omni
    naevo vindicatus: sive conatus geometricus quo
    stabiliuntur prima ipsa universae Geometriae Principia,
    was published in Milan in 1733. Saccheri missed his
    goal of providing an (indirect) proof of the postulate
    of parallels. But he did achieve a definitoric differen-
    tiation between geometries in light of the sum of the
    angles of the rectangle and also proved a series of
    theorems for the case of non-Euclidian (hyperbolic)
    geometry.

(2) In a letter to W. Bolyai dated June 3, 1882 Gauss
    wrote: "Bei einer vollständigen Durchführung (sc. einer
    Axiomatisierung der Geometrie, P.J.) müssen solche
    Worte wie 'zwischen' auch erst auf klare Begriffe
    gebracht werden, was sehr gut angeht, was ich aber
    nirgends geleistet finde." In a complete realization
    (sc., of an axiomatization of geometry - P.J.) such
    words as 'between' have to be conceptually clarified
    which is all very good but which I have never seen
    accomplished anywhere". (quoted from E. Kropp,
    Vorlesungen über Geschichte der Mathematik. Mannheim,
    1969, p. 35. See also C.F. Gauss, Werke. VOL. 8, p.222).

(3) The version of this conception widely held by presentday
    mathematicians can be found in H. Meschkowski's,
    Einführung in die moderne Mathematik (Mannheim: 1964,
    p.p. 15-19.) Of course, this version, according to
    which Hilbert's formalistic position is acknowledged
    to be correct (in opposition to G. Frege), not only
    sets out from false historical premises (see G. Gabriel,
    "implizite Definitionen - eine Verwechselungsgeschichte"
    to be published in the "Anals of Science"). It is also
    based upon a systematically false assessment of the
    controversy between Hilbert and Frege (see F. Kambartel,
    "Frege und die axiomatische Methode" in: Ch. Thiel
    (ed.), Frege und die moderne Grundlagenforschung.
    Meisenheim: 1975. p.p. 77-89).

(4) See E. Mach, Die Mechanik, historisch-kritisch darge-
    stellt, 9. Auflage, Leipzig 1933, p. 117.

(5) See G. Galileo: Discorsi e dimostrazioni matematiche, Firenze 1855, p. 199.

(6) G. Galileo, loc. cit., p. 201: "Ex quo pariter sequitur motum in horizontali esse quoque aeternum: si enim est aequablilis, non debilitatur, aut remittitur, et multo minus tollitur."

(7) Chr. Huygens, Tractatus de motu corporum ex percussione, 1703.

(8) Cf. B.E. Dijksterhuis, Die Mechanisierung des Weltbildes, Berlin, Göttingen, Heidelberg 1956, p. 387; E. Mach, loc. cit. p. 179; et al.

(9) G. Galileo, loc. cit., p. 149.

(10) G. Galileo, loc. cit., p. 172.

(11) Corpus omne perseverare in statu suo quiescendi vel movendi uniformiter in directum, nisi quatenus a viris impressis cogitur statum illum mutare.

(12) Cf. N.R. Hanson, Newton's First Law: A Philosophers Door into Natural Philosophy, in: R.G. Colodny (ed.), Beyond the Edge of Certainty, Prentice Hall 1965, p. 6-28, or: W. Stegmüller, Probleme und Resultate der Wissenschaftstheorie und Analytischen Philosophie, vol. II, 1, Berlin, Heidelberg, New York 1970, p. 110 f.

(13) Mutationen motus proportionalem esse vi motrici impressae, et fieri secundum lineam rectam qua vis illa imprimitur.

(14) Cf. footnote (12)

(15) Above all, E. Nagel; but also E. Mach.

(16) J.C. Sneed, The Logical Structure of Mathematical Physics. (Dordrecht: 1971).

(17) Stegmüller, Probleme und Resultate, VOL. II, 2., "Theorienstrukturen und Theoriendynamik". (Berlin, Heidelberg, New York: 1973. p. 119.)

(18) See W. K. Essler, Wissenschaftstheorie I, Definition
     und Reduktion. (Freiburg, München: 1970. p. 102):
     "That a concept is definable can only be decided in view
     of a given theory".

(19) See W. Stegmüller, loc. cit. p.p. 13, 119.

(20) Naturally, it is not presupposed here that one set out
     from a particular given axiomatization without
     preceding definitions.

(21) Quantitas Materiae est mensura ejusdem orta ex illius
     Densitate et Magnitudine conjunctim.

(22) ... Hanc autem quantitatem sub nomine corporis vel
     Massae in sequentibus passim intellego. Innotescit
     ea per corporis cujusque pondus.

(23) I shall not go into the concept of density that was
     taken up by Newton in association with the scholastic
     notion of 'quantitas materiae'. Here, density can,
     crudely speaking, be hypothetically characterized by the
     number of atoms per unitvolume. This concept of density
     does not open up any path towards making density or mass
     measureable.

(24) Such a requirement is to be understood concretely as the
     directive placed upon the manufacturer or else user
     of balance-scales to bring about symmetry and transitivity
     of t-equality.
     This is possible in a non-circular way, whereby such an
     assertion of possibility is, naturally, not founded by
     recourse to a mechanical theory, but rather by the
     explicit indication of procedures of production and
     application of balance-scales.

(25) For details, see, for instance, P. Janich, "Das Maß der
     Masse" in K. Lorenz (ed.): Konstruktionen versus
     Positionen. VOL. I (Berlin, New York: 1978).

(26) The two terms, 'central' and 'non-elastic', can be
     operatively determined in a non-circular way with the
     help of geometrical terms.

(27) Actioni contrariam semper et aequalem esse reactionem:
sive corporum duorum actiones in se mutuo semper esse
aequales et in partes contrarias dirigi.

A German version of this paper has already been
published in: Philosophia Naturalis 18 (1981), pp. 243-255.

INDEX

SYNTHESE LIBRARY

Studies in Epistemology, Logic, Methodology,
and Philosophy of Science

Managing Editor:
JAAKKO HINTIKKA (Florida State University, Tallahassee)

Editors:
DONALD DAVIDSON (University of California)
GABRIEL NUCHELMANS (University of Leyden)
WESLEY C. SALMON (University of Pittsburgh)

1. J. M. Bochénski, *A Precis of Mathematical Logic.* 1959.
2. P. L. Guiraud, *Problèmes et méthodes de la statistique linguistique.* 1960.
3. Hans Freudenthal (ed.), *The Concept and the Role of the Model in Mathematics and Natural and Social Sciences.* 1961.
4. Evert W. Beth, *Formal Methods. An Introduction to Symbolic Logic and the Study of Effective Operations in Arithmetic and Logic.* 1962.
5. B. H. Kazemier and D. Vuysje (eds.), *Logic and Language. Studies Dedicated to Professor Rudolf Carnap on the Occasion of His Seventieth Birthday.* 1962.
6. Marx W. Wartofsky (ed.), *Proceedings of the Boston Colloquium for the Philosophy of Science 1961-1962.* Boston Studies in the Philosophy of Science, Volume I. 1963.
7. A. A. Zinov'ev, *Philosophical Problems of Many-Valued Logic.* 1963.
8. Georges Gurvitch, *The Spectrum of Social Time.* 1964.
9. Paul Lorenzen, *Formal Logic.* 1965.
10. Robert S. Cohen and Marx W. Wartofsky (eds.), *In Honor of Philipp Frank.* Boston Studies in the Philosophy of Science, Volume II. 1965.
11. Evert W. Beth, *Mathematical Thought. An Introduction to the Philosophy of Mathematics.* 1965.
12. Evert W. Beth and Jean Piaget, *Mathematical Epistemology and Psychology.* 1966.
13. Guido Küng, *Ontology and the Logistic Analysis of Language. An Enquiry into the Contemporary Views on Universals.* 1967.
14. Robert S. Cohen and Marx W. Wartofsky (eds.), *Proceedings of the Boston Colloquium for the Philosophy of Science 1964-1966. In Memory of Norwood Russell Hanson.* Boston Studies in the Philosophy of Science, Volume III. 1967.
15. C. D. Broad, *Induction, Probability, and Causation. Selected Papers.* 1968.
16. Günther Patzig, *Aristotle's Theory of the Syllogism. A Logical-Philosophical Study of Book A of the Prior Analytics.* 1968.
17. Nicholas Rescher, *Topics in Philosophical Logic.* 1968.
18. Robert S. Cohen and Marx W. Wartofsky (eds.), *Proceedings of the Boston Colloquium for the Philosophy of Science 1966-1968.* Boston Studies in the Philosophy of Science, Volume IV. 1969.

19. Robert S. Cohen and Marx W. Wartofsky (eds.), *Proceedings of the Boston Colloquium for the Philosophy of Science 1966-1968.* Boston Studies in the Philosophy of Science, Volume V. 1969.

20. J. W. Davis, D. J. Hockney, and W. K. Wilson (eds.), *Philosophical Logic.* 1969.

21. D. Davidson and J. Hintikka (eds.), *Words and Objections. Essays on the Work of W. V. Quine.* 1969.

22. Patrick Suppes, *Studies in the Methodology and Foundations of Science. Selected Papers from 1911 to 1969.* 1969.

23. Jaakko Hintikka, *Models for Modalities. Selected Essays.* 1969.

24. Nicholas Rescher *et al.* (eds.), *Essays in Honor of Carl G. Hempel. A Tribute on the Occasion of His Sixty-Fifth Birthday.* 1969.

25. P. V. Tavanec (ed.), *Problems of the Logic of Scientific Knowledge.* 1969.

26. Marshall Swain (ed.), *Induction, Acceptance, and Rational Belief.* 1970.

27. Robert S. Cohen and Raymond J. Seeger (eds.), *Ernst Mach: Physicist and Philosopher.* Boston Studies in the Philosophy of Science, Volume VI. 1970.

28. Jaakko Hintikka and Patrick Suppes, *Information and Inference.* 1970.

29. Karel Lambert, *Philosophical Problems in Logic. Some Recent Developments.* 1970.

30. Rolf A. Eberle, *Nominalistic Systems.* 1970.

31. Paul Weingartner and Gerhard Zecha (eds.), *Induction, Physics, and Ethics.* 1970.

32. Evert W. Beth, *Aspects of Modern Logic.* 1970.

33. Risto Hilpinen (ed.), *Deontic Logic: Introductory and Systematic Readings.* 1971.

34. Jean-Louis Krivine, *Introduction to Axiomatic Set Theory.* 1971.

35. Joseph D. Sneed, *The Logical Sstructure of Mathematical Physics.* 1971.

36. Carl R. Kordig, *The Justification of Scientific Change.* 1971.

37. Milic Capek, *Bergson and Modern Physics.* Boston Studies in the Philosophy of Science, Volume VII. 1971.

38. Norwood Russell Hanson, *What I Do Not Believe, and Other Essays* (ed. by Stephen Toulmin and Harry Woolf). 1971.

39. Roger C. Buck and Robert S. Cohen (eds.), *PSA 1970. In Memory of Rudolf Carnap.* Boston Studies in the Philosophy of Science, Volume VIII. 1971.

40. Donald Davidson and Gilbert Harman (eds.), *Semantics of Natural Language.* 1972.

41. Yehoshua Bar-Hillel (ed.), *Pragmatics of Natural Languages.* 1971.

42. Sören Stenlund, *Combinators, λ-Terms and Proof Theory.* 1972.

43. Martin Strauss, *Modern Physics and Its Philosophy. Selected Papers in the Logic, History, and Philosophy of Science.* 1972.

44. Mario Bunge, *Method, Model and Matter.* 1973.

45. Mario Bunge, *Philosophy of Physics.* 1973.

46. A. A. Zinov'ev, *Foundations of the Logical Theory of Scientific Knowledge (Complex Logic).* (Revised and enlarged English edition with an appendix by G. A. Smirnov, E. A. Sidorenka, A. M. Fedina, and L. A. Bobrova.) Boston Studies in the Philosophy of Science, Volume IX. 1973.

47. Ladislav Tondl, *Scientific Procedures.* Boston Studies in the Philosophy of Science, Volume X. 1973.

48. Norwood Russell Hanson, *Constellations and Conjectures* (ed. by Willard C. Humphreys, Jr.). 1973.

49. K. J. J. Hintikka, J. M. E. Moravcsik, and P. Suppes (eds.), *Approaches to Natural Language*. 1973.
50. Mario Bunge (ed.), *Exact Philosophy – Problems, Tools, and Goals*. 1973.
51. Radu J. Bogdan and Ilkka Niiniluoto (eds.), *Logic, Language, and Probability*. 1973.
52. Glenn Pearce and Patrick Maynard (eds.), *Conceptual Change*. 1973.
53. Ilkka Niiniluoto and Raimo Tuomela,. *Theoretical Concepts and Hypothetico-Inductive Inference*. 1973.
54. Roland Fraissé, *Course of Mathematical Logic* – Volume 1: *Relation and Logical Formula*. 1973.
55. Adolf Grünbaum, *Philosophical Problems of Space and Time*. (Second, enlarged edition.) Boston Studies in the Philosophy of Science, Volume XII. 1973.
56. Patrick Suppes (ed.), *Space, Time, and Geometry*. 1973.
57. Hans Kelsen, *Essays in Legal and Moral Philosophy* (selected and introduced by Ota Weinberger). 1973.
58. R. J. Seeger and Robert S. Cohen (eds.), *Philosophical Foundations of Science*. Boston Studies in the Philosophy of Science, Volume XI. 1974.
59. Robert S. Cohen and Marx W. Wartofsky (eds.), *Logical and Epistemological Studies in Contemporary Physics*. Boston Studies in the Philosophy of Science, Volume XIII. 1973.
60. Robert S. Cohen and Marx W. Wartofsky (eds.), *Methodological and Historical Essays in the Natural and Social Sciences. Proceedings of the Boston Colloquium for the Philosophy of Science 1969-1972*. Boston Studies in the Philosophy of Science, Volume XIV. 1974.
61. Robert S. Cohen, J. J. Stachel, and Marx W. Wartofsky (eds.), *For Dirk Struik. Scientific, Historical and Political Essays in Honor of Dirk J. Struik*. Boston Studies in the Philosophy of Science, Volume XV. 1974.
62. Kazimierz Ajdukiewicz, *Pragmatic Logic* (transl. from the Polish by Olgierd Wojtasiewicz). 1974.
63. Sören Stenlund (ed.), *Logical Theory and Semantic Analysis. Essays Dedicated to Stig Kanger on His Fiftieth Birthday*. 1974.
64. Kenneth F. Schaffner and Robert S. Cohen (eds.), *Proceedings of the 1972 Biennial Meeting, Philosophy of Science Association*. Boston Studies in the Philosophy of Science, Volume XX. 1974.
65. Henry E. Kyburg, Jr., *The Logical Foundations of Statistical Inference*. 1974.
66. Marjorie Grene, *The Understanding of Nature. Essays in the Philosophy of Biology*. Boston Studies in the Philosophy of Science, Volume XXIII. 1974.
67. Jan M. Broekman, *Structuralism: Moscow, Prague, Paris*. 1974.
68. Norman Geschwind, *Selected Papers on Language and the Brain*. Boston Studies in the Philosophy of Science, Volume XVI. 1974.
69. Roland Fraissé, *Course of Mathematical Logic* – Volume 2: *Model Theory*. 1974.
70. Andrzej Grzegorczyk, *An Outline of Mathematical Logic. Fundamental Results and Notions Explained with All Details*. 1974.
71. Franz von Kutschera, *Philosophy of Language*. 1975.
72. Juha Manninen and Raimo Tuomela (eds.), *Essays on Explanation and Understanding. Studies in the Foundations of Humanities and Social Sciences*. 1976.

73. Jaakko Hintikka (ed.), *Rudolf Carnap, Logical Empiricist. Materials and Perspectives.* 1975.

74. Milic Capek (ed.), *The Concepts of Space and Time. Their Structure and Their Development.* Boston Studies in the Philosophy of Science, Volume XXII. 1976.

75. Jaakko Hintikka and Unto Remes, *The Method of Analysis. Its Geometrical Origin and Its General Significance.* Boston Studies in the Philosophy of Science, Volume XXV. 1974.

76. John Emery Murdoch and Edith Dudley Sylla, *The Cultural Context of Medieval Learning.* Boston Studies in the Philosophy of Science, Volume XXVI. 1975.

77. Stefan Amsterdamski, *Between Experience and Metaphysics. Philosophical Problems of the Evolution of Science.* Boston Studies in the Philosophy of Science, Volume XXXV. 1975.

78. Patrick Suppes (ed.), *Logic and Probability in Quantum Mechanics.* 1976.

79. Hermann von Helmholtz: *Epistemological Writings. The Paul Hertz/Moritz Schlick Centenary Edition of 1921 with Notes and Commentary by the Editors.* (Newly translated by Malcolm F. Lowe. Edited, with an Introduction and Bibliography, by Robert S. Cohen and Yehuda Elkana.) Boston Studies in the Philosophy of Science, Volume XXXVII. 1977.

80. Joseph Agassi, *Science in Flux.* Boston Studies in the Philosophy of Science, Volume XXVIII. 1975.

81. Sandra G. Harding (ed.), *Can Theories Be Refuted? Essays on the Duhem-Quine Thesis.* 1976.

82. Stefan Nowak, *Methodology of Sociological Research. General Problems.* 1977.

83. Jean Piaget, Jean-Blaise Grize, Alina Szeminska, and Vinh Bang, *Epistemology and Psychology of Functions.* 1977.

84. Marjorie Grene and Everett Mendelsohn (eds.), *Topics in the Philosophy of Biology.* Boston Studies in the Philosophy of Science, Volume XXVII. 1976.

85. E. Fischbein, *The Intuitive Sources of Probabilistic Thinking in Children.* 1975.

86. Ernest W. Adams, *The Logic of Conditionals. An Application of Probability to Deductive Logic.* 1975.

87. Marian Przelecki and Ryszard Wójcicki (eds.), *Twenty-Five Years of Logical Methodology in Poland.* 1977.

88. J. Topolski, *The Methodology of History.* 1976.

89. A. Kasher (ed.), *Language in Focus: Foundations, Methods and Systems. Essays Dedicated to Yehoshua Bar-Hillel.* Boston Studies in the Philosophy of Science, Volume XLIII. 1976.

90. Jaakko Hintikka, *The Intentions of Intentionality and Other New Models for Modalities.* 1975.

91. Wolfgang Stegmüller, *Collected Papers on Epistemology, Philosophy of Science and History of Philosophy.* 2 Volumes. 1977.

92. Dov M. Gabbay, *Investigations in Modal and Tense Logics with Applications to Problems in Philosophy and Linguistics.* 1976.

93. Radu J. Bogdan, *Local Induction.* 1976.

94. Stefan Nowak, *Understanding and Prediction. Essays in the Methodology of Social and Behavioral Theories.* 1976.

95. Peter Mittelstaedt, *Philosophical Problems of Modern Physics.* Boston Studies in the Philosophy of Science, Volume XVIII. 1976.

96. Gerald Holton and William Blanpied (eds.), *Science and Its Public: The Changing Relationship*. Boston Studies in the Philosophy of Science, Volume XXXIII. 1976.

97. Myles Brand and Douglas Walton (eds.), *Action Theory*. 1976.

98. Paul Gochet, *Outline of a Nominalist Theory of Proposition. An Essay in the Theory of Meaning*. 1980.

99. R. S. Cohen, P. K. Feyerabend, and M. W. Wartofsky (eds.), *Essays in Memory of Imre Lakatos*. Boston Studies in the Philosophy of Science, Volume XXXIX. 1976.

100. R. S. Cohen and J. J. Stachel (eds.), *Selected Papers of Léon Rosenfeld*. Boston Studies in the Philosophy of Science, Volume XXI. 1978.

101. R. S. Cohen, C. A. Hooker, A. C. Michalos, and J. W. van Evra (eds.), *PSA 1974: Proceedings of the 1974 Biennial Meeting of the Philosophy of Science Association*. Boston Studies in the Philosophy of Science, Volume XXXII. 1976.

102. Yehuda Fried and Joseph Agassi, *Paranoia: A Study in Diagnosis*. Boston Studies in the Philosophy of Science, Volume L. 1976.

103. Marian Przelecki, Klemens Szaniawski, and Ryszard Wójcicki (eds.), *Formal Methods in the Methodology of Empirical Sciences*. 1976.

104. John M. Vickers, *Belief and Probability*. 1976.

105. Kurt H. Wolff, *Surrender and Catch: Experience and Inquiry Today*. Boston Studies in the Philosophy of Science, Volume LI. 1976.

106. Karel Kosík, *Dialectics of the Concrete*. Boston Studies in the Philosophy of Science, Volume LII. 1976.

107. Nelson Goodman, *The Structure of Appearance*. (Third edition.) Boston Studies in the Philosophy of Science, Volume LIII. 1977.

108. Jerzy Giedymin (ed.), *Kazimierz Ajdukiewicz: The Scientific World-Perspective and Other Essays, 1931-1963*. 1978.

109. Robert L. Causey, *Unity of Science*. 1977.

110. Richard E. Grandy, *Advanced Logic for Applications*. 1977.

111. Robert P. McArthur, *Tense Logic*. 1976.

112. Lars Lindahl, *Position and Change. A Study in Law and Logic*. 1977.

113. Raimo Tuomela, *Dispositions*. 1978.

114 Herbert A. Simon, *Models of Discovery and Other Topics in the Methods of Science*. Boston Studies in the Philosophy of Science, Volume LIV. 1977.

115. Roger D. Rosenkrantz, *Inference, Method and Decision*. 1977.

116. Raimo Tuomela, *Human Action and Its Explanation. A Study on the Philosophical Foundations of Psychology*. 1977.

117. Morris Lazerowitz, *The Language of Philosophy. Freud and Wittgenstein*. Boston Studies in the Philosophy of Science, Volume LV. 1977.

118. Stanislaw Leśniewski, *Collected Works* (ed. by S. J. Surma, J. T. J. Srzednicki, and D. I. Barnett, with an annotated bibliography by V. Frederick Rickey). 1982. (Forthcoming.)

119. Jerzy Pelc, *Semiotics in Poland, 1894-1969*. 1978.

120. Ingmar Pörn, *Action Theory and Social Science. Some Formal Models*. 1977.

121. Joseph Margolis, *Persons and Minds. The Prospects of Nonreductive Materialism*. Boston Studies in the Philosophy of Science, Volume LVII. 1977.

122. Jaakko Hintikka, Ilkka Niiniluoto, and Esa Saarinen (eds.), *Essays on Mathematical and Philosophical Logic*. 1978.

123. Theo A. F. Kuipers, *Studies in Inductive Probability and Rational Expectation*. 1978.

124. Esa Saarinen, Risto Hilpinen, Ilkka Niiniluoto, and Merrill Provence Hintikka (eds.), *Essays in Honour of Jaakko Hintikka on the Occasion of His Fiftieth Birthday.* 1978.
125. Gerard Radnitzky and Gunnar Andersson (eds.), *Progress and Rationality in Science.* Boston Studies in the Philosophy of Science, Volume LVIII. 1978.
126. Peter Mittelstaedt, *Quantum Logic.* 1978.
127. Kenneth A. Bowen, *Model Theory for Modal Logic. Kripke Models for Modal Predicate Calculi.* 1978.
128. Howard Alexander Bursen, *Dismantling the Memory Machine. A Philosophical Investigation of Machine Theories of Memory.* 1978.
129. Marx W. Wartofsky, *Models: Representation and the Scientific Understanding.* Boston Studies in the Philosophy of Science, Volume XLVIII. 1979.
130. Don Ihde, *Technics and Praxis. A Philosophy of Technology.* Boston Studies in the Philosophy of Science, Volume XXIV. 1978.
131. Jerzy J. Wiatr (ed.), *Polish Essays in the Methodology of the Social Sciences.* Boston Studies in the Philosophy of Science, Volume XXIX. 1979.
132. Wesley C. Salmon (ed.), *Hans Reichenbach: Logical Empiricist.* 1979.
133. Peter Bieri, Rolf-P. Horstmann, and Lorenz Krüger (eds.), *Transcendental Arguments in Science. Essays in Epistemology.* 1979.
134. Mihailo Marković and Gajo Petrović (eds.), *Praxis. Yugoslav Essays in the Philosophy and Methodology of the Social Sciences.* Boston Studies in the Philosophy of Science, Volume XXXVI. 1979.
135. Ryszard Wójcicki, *Topics in the Formal Methodology of Empirical Sciences.* 1979.
136. Gerard Radnitzky and Gunnar Andersson (eds.), *The Structure and Development of Science.* Boston Studies in the Philosophy of Science, Volume LIX. 1979.
137. Judson Chambers Webb. *Mechanism, Mentalism, and Metamathematics. An Essay on Finitism.* 1980.
138. D. F. Gustafson and B. L. Tapscott (eds.), *Body, Mind, and Method. Essays in Honor of Virgil C. Aldrich.* 1979.
139. Leszek Nowak, *The Structure of Idealization. Towards a Systematic Interpretation of the Marxian Idea of Science.* 1979.
140. Chaim Perelman, *The New Rhetoric and the Humanities. Essays on Rhetoric and Its Applications.* 1979.
141. Wlodzimierz Rabinowicz, *Universalizability. A Study in Morals and Metaphysics.* 1979.
142. Chaim Perelman, *Justice, Law, and Argument. Essays on Moral and Legal Reasoning.* 1980.
143. Stig Kanger and Sven Öhman (eds.), *Philosophy and Grammar. Papers on the Occasion of the Quincentennial of Uppsala University.* 1981.
144. Tadeusz Pawlowski, *Concept Formation in the Humanities and the Social Sciences.* 1980.
145. Jaakko Hintikka, David Gruender, and Evandro Agazzi (eds.), *Theory Change, Ancient Axiomatics, and Galileo's Methodology. Proceedings of the 1978 Pisa Conference on the History and Philosophy of Science, Volume I.* 1981.
146. Jaakko Hintikka, David Gruender, and Evandro Agazzi (eds.), *Probabilistic Thinking, Thermodynamics, and the Interaction of the History and Philosophy of*

Science. Proceedings of the 1978 Pisa Conference on the History and Philosophy of Science, Volume II. 1981.

147. Uwe Mönnich (ed.), *Aspects of Philosophical Logic. Some Logical Forays into Central Notions of Linguistics and Philosophy.* 1981.

148. Dov M. Gabbay, *Semantical Investigations in Heyting's Intuitionistic Logic.* 1981.

149. Evandro Agazzi (ed.), *Modern Logic – A Survey. Historical, Philosophical, and Mathematical Aspects of Modern Logic and its Applications.* 1981.

150. A. F. Parker-Rhodes, *The Theory of Indistinguishables. A Search for Explanatory Principles below the Level of Physics.* 1981.

151. J. C. Pitt, *Pictures, Images, and Conceptual Change. An Analysis of Wilfrid Sellars' Philosophy of Science.* 1981.

152. R. Hilpinen (ed.), *New Studies in Deontic Logic. Norms, Actions, and the Foundations of Ethics.* 1981.

153. C. Dilworth, *Scientific Progress. A Study Concerning the Nature of the Relation Between Successive Scientific Theories.* 1981.

154. D. W. Smith and R. McIntyre, *Husserl and Intentionality. A Study of Mind, Meaning, and Language.* 1982.

155. R. J. Nelson, *The Logic of Mind.* 1982.

156. J. F. A. K. van Benthem, *The Logic of Time. A Model-Theoretic Investigation into the Varieties of Temporal Ontology, and Temporal Discourse.* 1982.

157. R. Swinburne (ed.), *Space, Time and Causality.* 1982.

158. R. D. Rozenkrantz, *E. T. Jaynes: Papers on Probability, Statistics and Statistical Physics.* 1983, forthcoming.

159. T. Chapman, *Time: A Philosophical Analysis.* 1982.

160. E. N. Zalta, *Abstract Objects. An Introduction to Axiomatic Metaphysics.* 1983, forthcoming.

161. S. Harding and M. B. Hintikka (eds.), *Discovering Reality. Feminist Perspectives on Epistemology, Metaphysics, Methodology, and Philosophy of Science.* 1983, forthcoming.

162. M. A. Stewart (ed.), *Law, Morality and Rights.* 1983, forthcoming.